ELECTRICAL ENGINEERING DEVELOPMENTS

ANALOG CIRCUITS

APPLICATIONS, DESIGN AND PERFORMANCE

ELECTRICAL ENGINEERING DEVELOPMENTS

Additional books in this series can be found on Nova's website
under the Series tab.

Additional E-books in this series can be found on Nova's website
under the E-book tab.

ELECTRICAL ENGINEERING DEVELOPMENTS

ANALOG CIRCUITS

APPLICATIONS, DESIGN AND PERFORMANCE

ESTEBAN TLELO-CUAUTLE
EDITOR

Nova Science Publishers, Inc.
New York

Copyright © 2012 by Nova Science Publishers, Inc.

All rights reserved. No part of this book may be reproduced, stored in a retrieval system or transmitted in any form or by any means: electronic, electrostatic, magnetic, tape, mechanical photocopying, recording or otherwise without the written permission of the Publisher.

For permission to use material from this book please contact us:
Telephone 631-231-7269; Fax 631-231-8175
Web Site: http://www.novapublishers.com

NOTICE TO THE READER

The Publisher has taken reasonable care in the preparation of this book, but makes no expressed or implied warranty of any kind and assumes no responsibility for any errors or omissions. No liability is assumed for incidental or consequential damages in connection with or arising out of information contained in this book. The Publisher shall not be liable for any special, consequential, or exemplary damages resulting, in whole or in part, from the readers' use of, or reliance upon, this material. Any parts of this book based on government reports are so indicated and copyright is claimed for those parts to the extent applicable to compilations of such works.

Independent verification should be sought for any data, advice or recommendations contained in this book. In addition, no responsibility is assumed by the publisher for any injury and/or damage to persons or property arising from any methods, products, instructions, ideas or otherwise contained in this publication.

This publication is designed to provide accurate and authoritative information with regard to the subject matter covered herein. It is sold with the clear understanding that the Publisher is not engaged in rendering legal or any other professional services. If legal or any other expert assistance is required, the services of a competent person should be sought. FROM A DECLARATION OF PARTICIPANTS JOINTLY ADOPTED BY A COMMITTEE OF THE AMERICAN BAR ASSOCIATION AND A COMMITTEE OF PUBLISHERS.

Additional color graphics may be available in the e-book version of this book.

LIBRARY OF CONGRESS CATALOGING-IN-PUBLICATION DATA

Analog circuits : applications, design, and performance / editor, Esteban Tlelo-Cuautle.
    p. cm.
 Includes bibliographical references and index.
 ISBN 978-1-61324-355-8 (hardcover)
 1. Electronic circuits. I. Tlelo-Cuautle, Esteban.
 TK7867.A5426 2011
 621.3815'3--dc22
               2011011509

Published by Nova Science Publishers, Inc. † New York

CONTENTS

Preface		**vii**
PART I: DESIGN		**1**
Chapter 1	Advances in Symbolic Techniques for Analog Design Automation *Guoyong Shi*	**3**
Chapter 2	Design Issues of SiGe HBT Based Analog Circuits *R.K. Chauhan*	**41**
Chapter 3	Approximation in Analog Signal Processing *J.M. David Báez-López*	**69**
PART II: APPLICATIONS		**91**
Chapter 4	Transconductance Amplifiers: NAM Realizations and Applications *Ahmed M. Soliman*	**93**
Chapter 5	Design of Current-Feedback Operational Amplifiers and Their Application to Chaos-Based Secure Communicatons *M.A. Duarte-Villaseñor, V.H. Carbajal-Gómez and E. Tlelo-Cuautle*	**121**
Chapter 6	Analog CMOS Morphological Edge Detector for Gray-scale Images *Luis Abraham Sánchez Gaspariano and Alejandro Díaz Sánchez*	**149**
PART III: PERFORMANCES		**169**
Chapter 7	Generalized Approach for Analog Network Optimization *A.M. Zemliak*	**171**
Chapter 8	A Technology-Aware Optimization of RF Integrated Inductors *P. Pereira, A. Sallem, M. H. Fino, M. Fakhfakh and F. Coito*	**213**

Chapter 9	Application of the ACO Technique to the Optimization of Analog Circuit Performances *B. Benhala, A. Ahaitouf, M. Kotti, M. Fakhfakh, B. Benlahbib, A. Mecheqrane, M. Loulou, F. Abdi and E. Abarkane*	**235**
Chapter 10	Analog Mismatch Analysis by Stochastic Nonlinear Macromodeling *Xue-Xin Liu, Yao Yu, Hai Wang and Sheldon X.-D. Tan*	**257**
Index		**275**

PREFACE

Up to now, the evolution of electronics has opened so many research areas. For instance, one can identify design, modeling and simulation, optimization and applications issues in analog, digital and mixed-signal circuits. Furthermore, this book presents recent developments and advances regarding the design, applications and performances of analog circuits. The first part includes three chapters focused on analog design automation and application of symbolic analysis, design issues for the future devices and circuits using silicon-germanium (SiGe) Heterojunction Bipolar Transistors (HBTs), and approximation in analog signal processing circuit design. The second part includes another three chapters presenting the application of transconductance amplifiers and realizations by applying the nodal admittance matrix technique, the automatic synthesis of current-feedback operational amplifiers and their applications to chaos-based secure communications, and application of amplifiers for the realization of an analog CMOS morphological edge detector for gray-scale images. The last part includes four chapters with the main goal to compute circuit performances through a generalized optimization approach, a technology-aware optimization technique for Radio Frequency integrated inductors, an application of the Ant Colonization Optimization technique, and an analog mismatch analysis by stochastic nonlinear macromodeling approach. Every chapter introduces challenges issues and discusses open lines for future scientific research.

Enjoy the Book!

PART I: DESIGN

In: Analog Circuits: Applications, Design and Performance   ISBN 978-1-61324-355-8
Editor: Esteban Tlelo-Cuautle   © 2012 Nova Science Publishers, Inc.

Chapter 1

ADVANCES IN SYMBOLIC TECHNIQUES FOR ANALOG DESIGN AUTOMATION

Guoyong Shi[*]
School of Microelectronics
Shanghai Jiao Tong University
Shanghai, China

Abstract

Remarkable advances have been made in the past decade on applying symbolic techniques for automating the design process of analog integrated circuits (ICs). Among all symbolic techniques, the technique of Binary Decision Diagram (BDD) has contributed the most. This chapter reviews the advanced BDD-based symbolic techniques developed in the past decade with special emphasis on their applications to automated analog circuit analysis. Design case studies are provided to demonstrate the usefulness of exact symbolic analysis of large operational amplifie circuits. A novel aspect of symbolic ac sensitivity computation is discussed with emphasis on its visualization. It is justifie via examples that the ac sensitivity plots are much easier to compute than directly calculating pole/zero sensitivities. Moreover, the graphical presentation of ac sensitivity can provide much more informative support for design aid. It is argued that the capability of a symbolic tool that can exactly analyze large analog circuit blocks is of great value in that a variety of design-oriented circuit analysis tasks can be performed automatically for the purpose of circuit optimization.

Keywords: alternate current (ac), analog integrated circuits, binary decision diagram (BDD), computer-aided design (CAD), determinant, graph, poles and zeros, sensitivity, symbolic analysis, transfer function

AMS Subject Classification 53D, 37C, 65P.

---

[*]E-mail address: shiguoyong@ic.sjtu.edu.cn. The author was supported by the National Natural Science Foundation of China (Grant No. 60876089).

1. Introduction

Operational amplifers (op-amps) are used in most analog integrated circuits (ICs), such as in switched capacitor flters, high-speed data converters, voltage references, and instrumentation amplifers, etc. The design techniques for a basic two-stage CMOS op-amp (shown in Fig. 1) were described in detail in the classical paper [1].

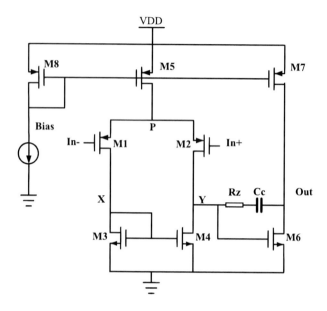

Figure 1. Two-stage operational amplifer (op-amp 1) with a nulling resistor R_z in series with a compensation capacitor C_c.

Important op-amp performance parameters to be considered include power dissipation, maximum allowable capacitive load, open-loop voltage gain, unity-gain frequency, output voltage swing, settling time, input ficker noise and thermal noise, power supply rejection ratio (PSRR), common-mode rejection ratio (CMRR), supply capacitance, and die area, etc. [1]. The pole-splitting compensation capacitor C_c and the nulling resistor R_z are important elements for shaping the frequency response. This design technique is still popular in the design practice today [2].

The publication of Gray and Meyer's work in 1982 was at the time of transition from bipolar technology to MOS technology; monolithic CMOS implementation of analog subsystem started to emerge [1]. However, from the analog design automation perspective, the design practice of the basic operational amplifers in the past thirty years has been lack of adequate support from design automation tools. The SPICE simulation tool is the dominating design aid. The design automation of analog integrated systems is further lagging behind the technology advancement.

The sophisticated design goals commonly posed in the design of a MOS op-amp as mentioned above indicate that developing a fully automated design tool for analog circuit design is inevitably more challenging than developing a numerical SPICE circuit simulator. Although a SPICE simulator can provide the one-way diagnosis of a topologically constructed and geometrically sized circuit, it does not provide any adequate information on how to tune the circuit toward a better performance. The designer must exercise his/her maximum intellectual resource in terms of experience and expertise to successfully arrive at a roughly optimized circuit.

Despite the great challenge lying ahead, research efforts for analog design automation have continued in the past decades. Symbolic techniques are among those receiving constant attentions from a limited number of researchers. The purpose of this chapter is to introduce one of the most promising symbolic techniques developed recently which makes use of the computation technique called *Binary Decision Diagram* (BDD). BDD was originally developed for digital circuit synthesis and verif cation [3]. The most prominent nature of BDD originates from its inherent property enforced by *sharing*. Data storage sharing is fundamental for a compact representation in many f elds, whether it is logic representation or sum-of-product (SOP) representation. Sharing can effectively suppress the computational complexity of many non-deterministic polynomial (NP) complete problems. Symbolic circuit analysis belongs to one class of such problems and is based on SOP representations in most formulations. Complexity suppression is of practical signif cance for analyzing much larger analog circuits by transforming a conventional analysis method into a BDD-based method.

Formulating a conventional circuit analysis method in the paradigm of BDD is in general not a trivial task. It requires both insights on problem formulation amenable to a BDD construction and specif c considerations in eff cient implementation. Some fundamental research efforts made in the past decade have not only proven the feasibility of applying BDD to symbolic analog circuit analysis, but also revealed great potentials that were not considered by the traditional symbolic techniques that largely incorporate approximation.

The signif cant research milestones in the past decade are reviewed in this chapter. The technical details are explained at such a level that those analog designers who are interested in the potential design automation tools would understand the underlying principle. This chapter is organized as follows. The advanced BDD-based symbolic methods are introduced in Section 2.. The basic principles used in two representative BDD-based symbolic analysis methods are explained via simple examples. Section 3. introduces a recently developed hierarchical analysis method built upon the two BDD methods. The main feature of this new method is its simplicity in problem formulation and its powerfulness in solving larger analog circuits. It is noteworthy that all the symbolic analysis methods surveyed in this chapter are *exact* methods in that no insignif cant terms are dropped in the f nal symbolic transfer functions. Because the symbolic network functions are exact, it is possible to carry out sensitivity analysis of the network function with respect to any circuit parameters, which leads to a notion called *symbolic ac sensitivity*. The usefulness of the symbolic ac sensitivity is discussed in Section 4.. If a symbolic network function is represented by a BDD, its sensitivity computation is fairly straightforward; that is, not much extra computational efforts are required. An unexpected byproduct of symbolic ac sensitivity computation is the exhibition of the poles-zeros dependence on the circuit parameters, which can be visualized in the frequency-domain plots. Examples presented in Section 5. demonstrate that the symbolic ac sensitivity can be used for multiple purposes, including understanding the dependence of circuit ac behavior on devices and certain quantitative metrics for device optimization. This chapter is concluded in Section 6..

2. Advanced Symbolic Methods

A nonlinear circuit can in general be described by the following differential equation

$$f\left(x(t), \frac{dx(t)}{dt}, u(t), t\right) = 0, \tag{1}$$

where $x(t)$ is the state of the circuit which describes the nodal voltages or branch currents introduced necessarily for circuit analysis and $u(t)$ stands for the externally applied voltage or current sources. A SPICE simulator can numerically solve the differential equation and provide transient analysis results given initial conditions.

Small-signal analysis is based on the linearization of nonlinear devices around a pre-characterized circuit operating point. Linearization of the nonlinear differential equation (1) would produce the following linear differential equation

$$C\frac{dx(t)}{dt} + Gx(t) = Fu(t), \tag{2}$$

where a time-invariant circuit is assumed. The construction of a small-signal model is mainly for the alternate current (ac) analysis when the circuit is excited by a small magnitude signal with varying frequency and phase components. Taking Laplace transform of the equation (2) produces the following algebraic equation

$$(Cs + G)X(s) = FU(s), \tag{3}$$

where s is the complex Laplace variable $s = \sigma + j\omega$. By choosing an appropriate output, one can write the output equation as

$$Y(s) = LX(s), \tag{4}$$

where L is a row vector. The input-output transfer function is therefore described by the following equation

$$Y(s) = L(Cs + G)^{-1}FU(s). \tag{5}$$

The coefficient of $U(s)$, defined by $H(s) := L(Cs + G)^{-1}F$, is the transfer function of the original circuit with the selected input and output.

The frequency-dependent gain and phase information is among the most valuable for analog designers. For example, shown in Fig. 2 is the ac response of the CMOS op-amp shown in Fig. 1 given certain sizing and biasing. The usefulness of the frequency response plots lies in the fact that the designer can easily read many frequency-domain design metrics from the plots, such as dc gain, bandwidth, unity-gain frequency, gain-margin, and phase-margin, etc., together with rough information on the distribution of the poles and zeros. Although such frequency response curves are useful in many aspects, they do not provide the designer any information on how to select one or several circuit devices for ac performance improvement or optimization. This is also the limitation of SPICE numerical simulation.

The above consideration is probably the key underlying thrust for the research on *symbolic* circuit analysis. Comparing to symbolically solving a nonlinear differential equation

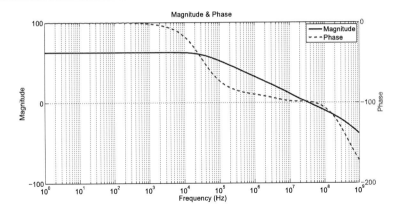

Figure 2. A frequency response example.

as given in (1) in the time-domain, symbolically solving a linearized equation in the frequency domain (or so called the *s-domain*) as given in (3) is tractable. Hence, over half a century, symbolic circuit analysis has mainly been studied in the ac domain, and its usefulness is widely recognized [4]. The key research problems in this area are the computational effciency problems and the development of tool functionality. The CAD researchers are responsible for providing easy to use, acceptable, and effective tool utilities for general analog designers.

A *symbolic* ac transfer function $H(s)$ can be more useful in many aspects than a numerical one. On the one hand, the ac plots as shown in Fig. 2 can be obtained repeatedly with much less computation by a symbolic transfer function (just by assigning new parameter values and evaluating the analytical formula). In this sense, one may select a circuit parameter and continuously change its value to visualize the interactively varying ac response curves in a graphical user interface, by which one can monitor how the bandwidth or the phase margin (or others) is being changed. Running SPICE for such a purpose is apparently less effcient and not convenient. On the other hand, it is also possible to derive other *interpretable* byproducts from a symbolic transfer function, such as simplifed transadmittances and input-output impedances, etc., as normally done by an analog designer. Some researchers even propose to derive approximate symbolic poles and zeros via symbolic analysis [5, 6]. However, due to the inherent computational diffculties involved, interpretable simplifcation and symbolic pole/zero extraction must use nonstandard and heavy postprocessing which carries quite an amount of tool developers' subjective judgement that might not be agreeable to circuit designers. For such reasons, symbolic CAD tools with such "advanced" functionalities have not become popular in the design community yet.

2.1. Basic Methods for Symbolic Analysis

We use the simple RC circuit shown in Fig. 3 to introduce some representative symbolic analysis methods that are still being used in more advanced algorithms. By inspection, the input-output voltage transfer function is found to be

$$V_{out} = \frac{1}{1+RCs} V_{in}. \tag{6}$$

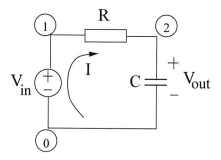

Figure 3. Simple RC circuit.

Formally, we can use either an algebraic method, which is used in numerical circuit simulators (such as SPICE), or a graphical method to derive the same symbolic solution in (6). The well-known algebraic method is the so called *modified nodal analysis* (MNA) method [7]. For the simple RC circuit, with the voltages at nodes '1' and '2' and the current f owing clockwise around the loop being the unknowns, we get the following linear equation

$$\begin{pmatrix} G & -G & -1 \\ -G & G+Cs & 0 \\ 1 & 0 & 0 \end{pmatrix} \begin{pmatrix} V_1 \\ V_{out} \\ I \end{pmatrix} = \begin{pmatrix} 0 \\ 0 \\ V_{in} \end{pmatrix}, \qquad (7)$$

where $G = 1/R$ is the conductance. Numerically, this equation can be solved by Gaussian elimination or the so-called LU factorization. Symbolically, this equation can be solved by Cramer's rule, which is written as follows:

$$V_{out} = \frac{N}{D} V_{in}, \qquad (8)$$

where N and D are two determinants given below

$$N = \begin{vmatrix} G & 0 & -1 \\ -G & 0 & 0 \\ 1 & 1 & 0 \end{vmatrix}, \quad D = \begin{vmatrix} G & -G & -1 \\ -G & G+Cs & 0 \\ 1 & 0 & 0 \end{vmatrix}. \qquad (9)$$

It is easily verif ed that the quotient of the two determinants is equal to the transfer function given in (6).

The same circuit problem also can be solved graphically (or topologically) by enumerating certain spanning trees of a given graph. Those terms generated from the spanning trees form the transfer function. There are a variety of ways to enumerate spanning trees for the purpose of network function generation. A recently developed rule-based enumeration method [8] can produce the spanning trees shown in Fig. 4, from which three product terms are generated, two from the two spanning trees and one from the pair of spanning trees.

According to the rules derived in [8], the input-output (I/O) pair is modeled as a controlled source. For the example of voltage input and voltage output, the I/O pair is treated as a *voltage controlled voltage source* (VCVS), with the output being the *controlling* voltage source (VC) and the input being the *controlled* voltage source (VS). The reason for such a treatment is explained in [8]. The novel aspects of the enumeration rules that differentiate

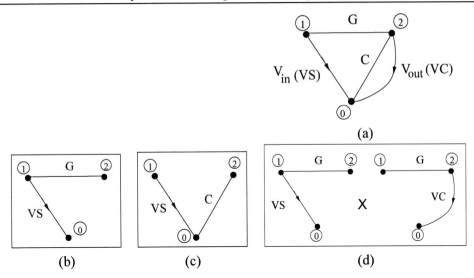

Figure 4. Graph and spanning trees for the simple RC circuit: (a) The graph for spanning tree enumeration; (b) Tree 1 generates the term G; (c) Tree 2 generates the term C; and (d) Tree pair generates the term $-GX$.

them from those classic spanning tree approaches, such as the *two-tree* method used in [9], are in the treatment of the four dependent sources. Those classic methods can easily deal with VCCS but have problems with the other dependent sources VCVS, CCCS, and CCVS in that extra conversions must be made. In contrast, the rules developed in [8] can equally deal with all four dependent sources without the extra need of conversion.

The rules established in [8] are literally lengthy to state (hence are not repeated here), but they are implementable by modern computer programming [10]. The three spanning tree objects shown in Fig. 4 would generate the following three *signed* product terms summing up to zero, i.e.,

$$G + Cs - GX = 0, \tag{10}$$

where X represents the coefficient of the VCVS pair, namely, $V_{in} = X \cdot V_{out}$. Equation (10) leads to

$$\frac{V_{out}}{V_{in}} = X^{-1} = \frac{G}{G + Cs}, \tag{11}$$

which is the same voltage transfer function derived in (6) for the simple RC circuit in Fig. 3.

The basic principles involved in the above two representative methods, one *algebraic* and the other *topological*, are not radically new; they have been studied in a variety of forms in the literature since 1950's. The monograph by P. M. Lin [11] published in 1991 presents a comprehensive survey on the those traditional methodologies for symbolic circuit analysis. Other overview works on this subject are [4, 12]. Although those classical methods have been well studied, they have not been able to grow into industrial design tools. The main reason is that those methods are all of the nature of enumeration, hence of exponential complexity, not amenable to analyzing large analog integrated circuits. All symbolic methods that claimed to be able to solve large analog circuits before 2000 were based on *approximate* analysis. When the Binary Decision Diagram (BDD) was introduced to the area of symbolic circuit analysis around 2000, the circuit size that could be exactly analyzed by a symbolic method was increased greatly.

2.2. Binary Decision Diagrams for Symbolic Analysis

The two representative classes of methods outlined in the preceding subsection would not be of practical use without using a superior data structure to manipulate large determinants and large graphs. Here, by "large" we mean that the number of circuit nodes could be in the range of 20 to 40 or more, which can potentially meet the practical design needs.

The past ten years of research have validated that the Binary Decision Diagram data structure proposed by Randal Bryant in 1986 [3] can be used as an eff cient data structure for not only a compact representation of symbolic transfer functions, but also a means of eff cient data manipulation and evaluation.

BDD was originally developed for eff cient representation of logic functions, which are processed traditionally in the form of sum-of-products (SOP) or product-of-sums (POS) expressions. When the number of literals reaches certain level, the count of terms appearing in such expressions would not be favorable for eff cient computer processing. More critically, the SOP or POS expressions are not *canonical* in that multiple literally different boolean expressions exist for a logic function def ned by the same truth table. Bryant's fundamental contribution in his 1986 paper [3] was the establishment of BDD *canonicity* that exists with an *ordered* list of symbols. Although there exists certain symbol order that can make a BDD as compact as possible, f nding an optimal order for such compactness is inherently an NP-complete problem. Hence, heuristics are used in practice. While the canonicity of BDD is important for applications in logic synthesis and formal verif cation [13], the data compactness achieved by *sharing*, which is the underlying mechanism contributing to the canonicity of BDD, can be a powerful means for representing symbolic transfer functions in analog circuit analysis. Moreover, analytical operations (such as functional compositions and derivatives) can be implemented very easily and eff ciently based on a BDD.

The application of BDD to symbolic circuit analysis emanates from a simple observation of the similarity between the SOP representations of logic circuits and the algebraic SOP expressions arising from the determinant based algebraic method or the spanning tree based topological method for analog circuits. The most crucial question for a successful application is: *How to construct a BDD-based transfer function given a linearized circuit?*

Over the years, many BDD packages have been developed in the research area of logic synthesis. Certain form of standardization has been reached. For example, it is not necessary to develop BDD operations for different logic operations such as *AND*, *OR*, and *XOR*, etc. One may simply use one single operation called *If-Then-Else* (ITE) to handle all different logic operations [14]. Logic operations also can be formulated in sets, which leads to set-based BDDs [15]. Set-based BDDs are more general so that algebraic SOP expressions are included as special cases. The f rst application of using a set-based BDD for representing the product terms of determinant expansion was realized by Shi and Tan in [16]. Later on it was demonstrated that determinant expansion also can be treated as a logic synthesis process [17], for which one may directly use a logic BDD package. However, these approaches do not have much difference in the underlying implementation mechanism, therefore, their eff ciency does not make big difference.

A novel perspective is not to formulate analog circuit problems logically, rather to treat them directly in their underlying formulation, whether algebraic or topological. If in an algebraic formulation, determinants are the objects for BDD sharing, whereas in a topological formulation, graphical objects are used for BDD sharing. BDD construction in such a perspective is fundamentally different from its construction in logic synthesis, but the BDD sharing mechanism and canonicity are retained, which contributes the major eff ciency to symbolic analysis of analog circuits. The introduction in the following subsections takes the new perspective for BDD construction.

2.3. Determinant Decision Diagram

Shi and Tan [16] contributed the pioneering work of applying BDD to symbolic circuit analysis. In that work, they developed algorithms for expanding a determinant using a set-based BDD package (called *Zero-suppressed BDD*, ZBDD) [15, 18]. As stated earlier, by Cramer's rule, the I/O transfer function can be expressed as the quotient of two determinants. So, as long as the expansions of the two determinants can be constructed in one BDD (shared), a symbolic representation of the transfer function is constructed in the computer memory. The authors of [16] named such a data structure *Determinant Decision Diagram* (DDD).

The "sharing" mechanism in BDD can be implemented in computer programming by a *hash table*. The design of hash scheme is nonstandard and implementation dependent. However, the eff ciency of a BDD-based symbolic package largely depends on the implementation strategy of hash scheme and hash table organization.

Two operations must be def ned before formulating a problem in the form of BDD construction. In the case of a determinant, the operations of "*Minor*" and "*Remainder*" are def ned with respect to any selected matrix element [16]. Here, by "*Minor*" it refers to the operation of deleting a row and a column intersected at the selected element, while by "*Remainder*" it refers to setting the selected element to *zero*. This can be brief y expressed by the following equation

$$\det A = (-1)^{i+j} a_{i,j} \cdot Minor(A, a_{i,j}) + Remainder(A, a_{i,j}), \tag{12}$$

where the binary decision is made at the ith row and jth column element a_{ij} of matrix A. Equation (12) is a valid identity in linear algebra. Assume that A is an $n \times n$ matrix. Then in equation (12) $Minor(A, a_{i,j})$ is a reduced dimensional determinant of dimension $(n-1)$ while $Remainder(A, a_{i,j})$ is still a determinant of dimension n. As long as both determinants $Minor(A, a_{i,j})$ and $Remainder(A, a_{i,j})$ are nonsingular, the binary expansion of (12) can be repeated until the f nal determinants are either *scalar* or *singular*.

Two things crucial to the expansion process are: *order of expansion* and *minor sharing*, which make a DDD canonical [16]. The *order of expansion* refers to the notion of symbol ordering in BDD; one has to choose either an explicit (pre-chosen) order or an implicit (runtime) order for sequencing the determinant expansion. A pre-chosen order requires extra computation time, while a runtime order does not. The role of *minor sharing* is more important, because without sharing the minors, the expansion of (12) is nothing else than a binary expansion, it requires rapidly growing computer memory with a large exponential base factor.

The possibility of *sharing* in determinant expansion can be illustrated by the following example. Suppose we have a 4×4 determinant

$$\begin{vmatrix} a_{11} & a_{12} & a_{13} & a_{14} \\ a_{21} & a_{22} & a_{23} & a_{24} \\ a_{31} & a_{32} & a_{33} & a_{34} \\ a_{41} & a_{42} & a_{43} & a_{44} \end{vmatrix}. \tag{13}$$

The terms in the expansion of the lower-right 2×2 determinant

$$\begin{vmatrix} a_{33} & a_{34} \\ a_{43} & a_{44} \end{vmatrix} = a_{33}a_{44} - a_{34}a_{43} \tag{14}$$

can be shared in the following sense: the two successive *Minor* operations on a_{11} followed by a_{22} arrive at minor (14) while applying the other two successive *Minor* operations on a_{12} followed by a_{21} also arrives at that minor. Therefore, the 2×2 minor in (14) only has to be expanded once in the expansion of the original full determinant.

The design of a hash scheme should take into account of the features of the objects under operation. Both the determinant objects in the algebraic approach and the graph objects in the topological approach are the candidate objects for hash operations. The critical components in the procedure of implementing a BDD-based symbolic package are discussed next.

Since the main sharable objects in determinant expansion are those sub-dimensional minors, the key issue one should consider is how to hash (or look up) the minors in the speediest way. The technical details in implementation can make considerable difference in runtime eff ciency. The original implementation strategy presented in [16] did not make use of a good property inherent in the *ordered* determinant expansion. The property discovered later in [19] is stated here:

Property of Minor Uniqueness [19]

Given an expansion order, the row and column indices of minor are sufficient for uniquely identifying a minor, regardless of the minor entries.

A hash scheme can be implemented based on this property. The advantage is that a logic or set-based BDD package is not needed anymore to realize sharing, as discussed in greater detail in the recent paper [20].

With the advancement achieved by the principle of DDD and some recent improvements, larger analog circuit blocks such as op-amp $\mu A741$ and $\mu A725$ containing about 20 to 26 bipolar transistors can be analyzed in *exact* symbolic expressions, without introducing any approximation. Exact symbolic analysis has many applications, in particular, the symbolic sensitivity analysis requires exact symbolic expressions. We shall discuss in further detail later that symbolic sensitivity analysis can provide much more important design information than that symbolic network functions only can provide.

Although the DDD method signif cantly improves the capacity of symbolic analysis, this improvement still has limit. Specif cally, it would be hard for DDD to derive exact symbolic transfer functions for circuits larger than $\mu A741$ and $\mu A725$. More sophisticated techniques (such as hierarchy) must be introduced into DDD to further improve its capacity.

A number of extensions based on the DDD algorithm have been published in the literature, such as [17, 21–25], among others. These later extensions have pursued explorations

Advances in Symbolic Techniques for Analog Design Automation 13

in a variety of directions, including matrix-based hierarchical analysis methods and a logic synthesis approach to DDD construction. But the following limitations remain: 1) No great improvement on the analyzable circuit size has been reported; for example, for op-amps containing more than 40 MOSFETs. 2) The theoretical aspects of complexity suppression achievable by using a DDD have never been investigated seriously until the recent work of [19], where a preliminary but interesting answer is provided for full matrices. 3) It is still not well understood what tool functionalities should be developed based on exact symbolic analysis results. Some of the limitations will be discussed further in the later sections.

2.4. Graph-Pair Decision Diagram

The success of DDD as an application of BDD for symbolic analysis of analog circuits has inspired a new direction of research. A parallel extension would be to apply BDD for representing the process of spanning-tree enumeration. Just like determinant expansion, spanning-tree enumeration also can be carried out in a *non-exhaustive* way. For this purpose, appropriate binary decisions must be defned in such a way that BDD sharing can be incorporated in the enumeration process.

Applying BDD to spanning-tree enumeration f rst appeared in the work of [8,10] around 2007. Enumerating all spanning-trees of a connected graph was considered by Minty in his one page paper [26] published in 1965, where Minty invented a *binary* graph decomposition process by applying one of the operations "*In*" and "*Out*" to the ordered sequence of edges, where the "*In*" operation retains the edge and the "*Out*" operation removes the edge. Figure 5 shows an example of using the Minty algorithm to enumerate the f ve spanning trees of a graph containing four edges. The operations follow the edge ordered by $e_1 < e_2 < e_3 < e_4$, where '$<$' reads *precedes*. A checking mechanism is needed to monitor whether a spanning tree has formed or whether the edges left are insuff cient for forming a spanning-tree. It is apparent that the Minty algorithm is a binary process that enumerates all spanning trees exhaustively, which are $T_1(e_1, e_2)$, $T_2(e_1, e_3)$, $T_3(e_1, e_4)$, $T_4(e_2, e_3)$, and $T_5(e_2, e_4)$ for the given example.

In 1965 Minty was not aware of sharing the subgraphs generated in the decomposition process. What Minty obtained was an exhaustive binary decomposition procedure, whose exponential complexity does not allow the algorithm for solving large graphs.

Considering from the perspective of BDD, the Minty algorithm can be modif ed slightly to incorporate the *sharing* in the process of binary decomposition. The modif cation lies in the way the subgraphs are produced. In order to take the advantage of sharing, the subgraphs in the process of decomposition should be made comparable for isomorphism. A good way of achieving such a property is by *graph reduction* instead of *retaining/removing* edges. More specif cally, when an edge is to be retained, one should *collapse* the edge to one node instead of retaining the edge, while an edge to be removed is simply *removed*. In this sense, the graph operations are still binary, but become *collapsing* and *removal*. In this way, the original graph is gradually reduced with the numbers of edges in the subgraphs decreasing successively until one of the following cases is encountered: (i) only one node is left, (ii) the remaining edges are insuff cient for completing a spanning-tree, or (iii) the subgraph becomes disconnected.

Figure 6 illustrates the application of the modif ed Minty algorithm to the same graph

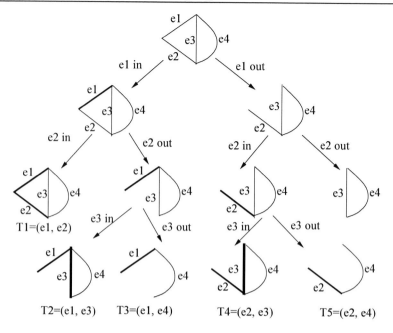

Figure 5. Minty algorithm for enumerating all spanning trees of a four-edge graph.

appeared in Fig. 5. We see that the middle subgraph in the third row (containing edges e_3 and e_4) is shared by two preceding subgraphs in the second row. Note that, without collapsing the edges e_1 and e_2, the two subgraphs in the middle of the third row in Fig. 5 are not identical. Since one binary decision is always applied to a selected edge, the name of that edge can be the name of a BDD vertex. Also note that the three terminations marked by "1" in Fig. 6 are essentially a degenerated subgraph containing a single node, they can share one single BDD vertex to indicate the termination of spanning-trees. Putting these considerations together we arrive at the Binary Decision Diagram shown in Fig. 7. The subgraph attached to each BDD vertex illustrates the subgraph under operation there.

Note that we use *solid* arrows in Fig. 7 for the "*In*" operations and *dashed* arrows for the "*Out*" operations. Therefore, the spanning trees represented in the BDD can be read out as follows. Starting from the root vertex, we traverse downward to the terminal vertex "1". If a solid arrow emerges from a vertex, the vertex name (i.e., the edge) becomes an edge of the spanning tree we are traversing. If a dashed arrow emerges from a vertex, that edge is excluded from the spanning tree. There exist f ve paths starting from the root vertex and arriving at the terminal vertex "1". Therefore, f ve spanning trees result by the stated rule (see that listed in Fig. 7.) In this sense, we say that all the spanning trees for the given graph are *implicitly* represented by the constructed BDD.

Similarly to the determinant expansion where reduced dimensional minors obtained by the operations of "*Minor*" and "*Remainder*" f nd sharing among themselves, the partially reduced subgraphs generated by the operations of edge "*In*" (i.e., *collapsing*) and edge "*Out*" (i.e., *removal*) also f nd sharing among themselves. It is possible to devise a graph comparison mechanism to identify the isomorphism between two equal-dimensional subgraphs [10].

Applying the above idea to symbolic circuit analysis requires further work. As we

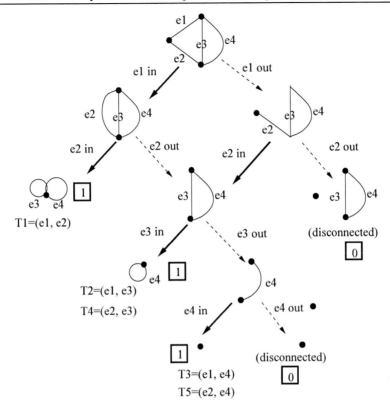

Figure 6. Minty algorithm improved for sharing the subgraphs. The edges forming loops can be removed.

mentioned earlier, the topological method for solving the simple RC circuit example has to deal with dependent sources. When dependent sources exist in a circuit, a graphical formulation has to take into consideration of the dependence between edges. Therefore, enumerating spanning-trees only is not sufficient for enumerating the symbolic product terms. Instead, more sophisticated enumeration rules must be developed to take care of the edge dependence. The two-graph method [9] is one of such examples where two graphs are constructed for dealing with circuits containing dependent sources. However, the two-graph method has limitation in that it is only applicable to the type of VCCS dependent sources, not all four types of dependent sources (see [9]).

The work by Shi et al. [8] developed a new set of tree-pair enumeration rules that can be applied to all four types of dependent sources. Moreover, the established rules can easily be adapted to graph reduction operations just as we developed for the modified Minty algorithm. In the new graphical formulation, a notion called *pair of spanning-trees* (or simply called *tree-pairs*) is introduced, which is analogous to the *two-graph* method. Because the dependent sources introduce constraints to the interrelation between edges, not all spanning tree-pairs are valid for a symbolic transfer function. Only those satisfying certain conditions are admissible and have to be enumerated [8]. Also, to facilitate the enumeration of admissible spanning tree-pairs, an initial pair of graphs must be constructed.

After establishing the enumeration rules, the next step is to convert explicit tree-pair

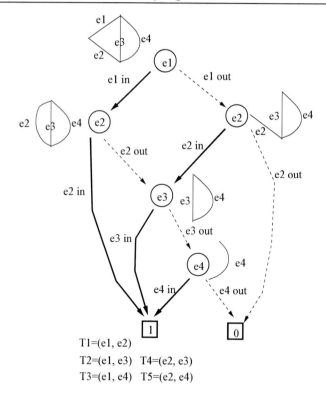

Figure 7. Five spanning trees represented in a BDD.

enumeration into an implicit enumeration process by embedding the enumeration rules in the process of a BDD construction. In this process, the same strategy introduced for the modifed Minty algorithm is adopted for reducing a pair of graphs; that is, edge retaining is treated by edge collapsing. This strategy again makes it possible to share the reduced dimensional subgraph pairs. The enumeration rules are enforced at each step of binary decisions. The resulting binary decision diagram is termed *Graph-Pair Decision Diagram* (GPDD) [8,10]. The implementation details are presented in [10] with experimental results. It was demonstrated that the GPDD method could also solve those circuits such as $\mu A741$ and $\mu A725$ once solved by DDD. The memory and time effciency was comparable to that of DDD, sometimes GPDD could be better because the two BDD construction methods use different sets of symbols and hashing objects, hence symbol ordering heuristics are completely different.

In summary, two BDD-based symbolic algorithms have been developed in the past decade, with radically different problem formulations. One might ask what the critical difference is between the two algorithms. The most critical difference in the author's opinion is in the defnition of symbols and its consequences. In DDD, the MNA matrix elements are the symbols, but they are not directly the circuit elements, as seen from the example matrix in (7). For example, the (2, 2) entry of the coeffcient matrix is $a_{22} = G + Cs$, which is treated as a single symbol in DDD. Also the same circuit element G appears in the multiple positions of a_{11}, a_{12}, and a_{21} and they are treated as different symbols in DDD. Consequently, cancellation is inevitable in the formulation of DDD. In contrast, in the GPDD formulation the symbols are just the circuit elements and the enumeration rules preclude cancellation [8, 10]. In modern computing, cancellation might not be a very serious problem, but the different symbol def nitions do lead to differences in the implementation of sensitivity analysis. Also, a hierarchical analysis strategy can take the advantage of topological formulation of GPDD to achieve the maximal amount of substructure sharing, which is to be discussed in the next section.

3. Symbolic Stamps for Hierarchical Analysis

It is important to be aware of how a BDD improves the eff ciency for those exponentially complex problems. It is the *sharing* mechanism enforced in the BDD construction that helps reduce the enumeration complexity. Whenever a substructure is shared, the detailed operations on the substructure are performed *only once*, rather than being repeatedly executed in those exhaustive enumeration methods. This fact must be observed in all successful applications of BDD.

However, it is also important to be aware that the application of a BDD does not change the nature of an exponentially complex problem; that is, a problem with exponential complexity would in general remain exponentially complex regardless of whether or not a BDD is used. What BDD can contribute is the possibility of reducing the base factor of exponential increase. For some reason this fact has not been well understood for a long time until recently the work [19] showed that the optimal DDD size (i.e., the number of BDD vertices) for any $n \times n$ full matrix (without any zero element) is $n2^{n-1}$. Obviously, the optimal complexity is still an exponential function of the problem size, but the base is much lower than the counterpart of explicit determinant expansion. Reducing the exponential base factor simply means that larger problems can become tractable by means of a BDD reformulation. However, It also implies that the limit of problem size still exists.

The merits of the BDD-formulated methods lie not only in the increased capability of solving larger problems, but also in the remarkable eff ciency gain in numerical evaluation, which must be performed repeatedly in some circuit analysis tasks. However, both of the BDD-formulated methods, DDD and GPDD, could not solve op-amp circuits containing more than 30 bipolar or MOS devices in exact manner. Therefore, hierarchical methods have been introduced to the BDD-formulated algorithms.

Two hierarchical schemes built on the top of DDD have been proposed in [21] and [24]. These methods essentially are based on matrix partitioning and factorization of block matrices. The structural circuit information is not effectively used in the hierarchical formulation. These efforts did achieve certain capacity improvement in problem size, but were not able to substantially improve the DDD capability for solving circuits containing more than 30 MOSFETs. A substantial advancement was made in the recent work [27], where attempt was made to adopt a simple yet effective idea called *symbolic stamps*.

3.1. Symbolic Stamp Method

Shown in Fig. 8 is a commonly used MOS small-signal model, which is used in SPICE ac analysis [28]. In symbolic circuit analysis, all semiconductor devices are substituted by their small-signal models; the small-signal element names appear as symbols in the

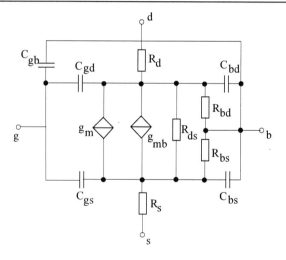

Figure 8. MOS level 3 small signal model [28].

analytical transfer functions. A typical op-amp circuit usually contains 10 to 50 (or more) semiconductor devices. Directly substituting all bipolar or MOS devices by their small-signal models would increase the size of MNA matrix or the size of graph, which in turn would increase the solving complexity of f at (non-hierarchical) symbolic methods.

According to the principle of circuit simulation, a stamp is created for a device (e.g., a MOSFET) in the MNA matrix, where all other device stamps are assembled [29]. Symbolically speaking, all stamps for the same type of devices are identical, except that the element values may change depending on the operating condition of each device. This fact can be utilized in the scenario of symbolic analysis. We just create one symbolic stamp for one type of devices like MOSFET and use it as a calculator; passing the specif c small-signal device values to the symbolic stamp, numerical stamp values for that device are obtained. This treatment is meaningful in the symbolic setting in that the computational complexity for large integrated circuits can be reduced.

For an m-port device like a MOSFET, its symbolic stamp is an $m \times m$ transadmittance matrix consisting of m^2 transadmittances. They are essentially transfer functions from the voltage at the ith port to the current at the jth port for $i, j = 1, \cdots, m$. One may run a DDD or GPDD package for m^2 times to derive m^2 BDDs for all transadmittances in the symbolic stamp. But the drawback is that independently deriving m^2 transadmittances would sacrif ce quite an amount of eff ciency, hence that strategy is not favorable.

The work [27] made an attempt of using a *multi-root* GPDD for the symbolic stamp, in which the m^2 transadmittances are constructed by sharing one single GPDD. The rationale is that the m^2 transadmittances, although distinct, are transfer functions of the same circuit structure. Hence, the subgraphs involved in the GPDD construction must be sharable in high proportion. It is expected that such a strategy can lead to a much more compact representation of the symbolic stamp, benef cial to both the storage and evaluation eff ciency.

In [8], a *voltage-input current-output* transadmittance is treated as a CCVS (rather than a VCCS). The graph reduction method employed there constructs a *multi-root* GPDD for the m^2 initial graph pairs. Those m^2 initial graph pairs closely resemble each other except for the def nition of input-output ports. Hence, further simplif cations are possible in the construction of the multi-root GPDD, which justif es a serious study on the implementation details (see [8]). We stress that the resulting GPDD is an m^2-root GPDD, each GPDD root providing an access to one transadmittance. Another reason for choosing a GPDD for the symbolic stamp construction is due to an important property coming with a GPDD; the GPDD symbols are directly the circuit elements. This property is of special signif cance when such a symbolic tool is used for applications such as MOS device *sizing* where direct device-parameter-to-symbol mapping provides convenience.

For the four-port small-signal circuit shown in Fig. 8, its symbolic stamp has 16 entries. The corresponding 16-root GPDD can be derived very quickly in less than one second. In our experiment, by using an arbitrary symbol order the constructed 16-root GPDD contained 481 BDD vertices. Running one numerical evaluation of the symbolic stamp represented by the GPDD also takes inappreciable time.

As the four-port small-signal model is represented by a 4×4 symbolic stamp, the internal nodes of the small-signal subcircuit would not appear in the MNA matrix. Consequently, the MNA matrix size is smaller than a f at formulation. Since the computational complexity of symbolic analysis is inherently exponential, reducing the problem size even a little can greatly reduce the computational complexity. This implies that if we use a DDD to solve a reduced MNA equation assembled in symbolic stamps, the DDD complexity can be reduced as well.

Fig. 9 shows the hierarchical strategy we have described so far, in which a DDD solver lies at the top level for solving the assembled MNA matrix with the symbolic stamps constructed by the multi-root GPDD solver lying at the bottom level. If more than one type of devices is present in a circuit, each type of devices can be handled by one multi-root GPDD.

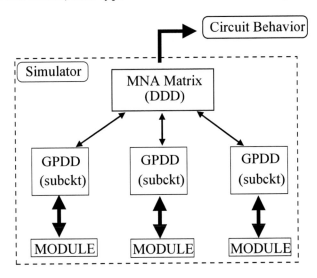

Figure 9. Hierarchical analysis structure.

The hierarchical symbolic analysis procedure using symbolic stamps is summarized as follows.

Hierarchical Analysis Procedure [27]:

Step 1. Choose a small-signal model for the transistors (devices) used in the circuit.

Step 2. Construct a multi-root GPDD for each multiport symbolic stamp.

Step 3. Formulate an MNA matrix and create a device table for stamp evaluation.

Step 4. Run DDD to construct a symbolic transfer function.

Step 5. Run hierarchical numerical evaluations.

3.2. Experimental Results

A C++ symbolic simulator called *HybridSim* has been implemented in the work [27] based on the hierarchical scheme using symbolic stamps. The test data were collected from a desktop computer installed with an AMD Athlon64 2.20GHz processor and 2GB memory. The following two benchmark circuits were analyzed:

- Op-amp 2: A two-stage Miller MOSFET amplifier (with a folded-cascode first stage), containing 24 transistors (Fig. 10).

- Op-amp 3: A MOSFET amplifier containing 44 transistors (Fig. 11).

The level-3 small-signal model (containing 12 symbols) given in Fig. 8 is used for all MOS transistors in both amplifier circuits.

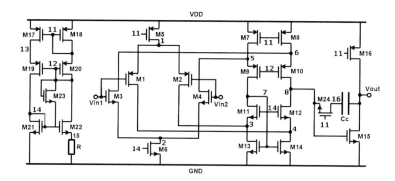

Figure 10. Op-amp 2: A rail-to-rail Miller MOS op-amp containing 24 transistors.

Summarized in Table 1 are the experimental results. Listed in the column "#Symb (GPDD)" is the total number of symbols in the small-signal subcircuit. Listed in the column "#Symb (DDD)" are the numbers of nonzero elements in the MNA matrices. The MNA matrix sizes are listed in the column "Matrix Size", which are the problem sizes handled by the DDD solver. The column $|GPDD|$ lists the number of GPDD vertices created for the small-signal model, while the column $|DDD|$ lists the numbers of DDD vertices created for the MNA matrices. The column "Time" lists the total simulation time used for analyzing each circuit, which includes circuit parsing time, the construction time for two layers of decision diagrams, and the time for running one round of ac analysis with 100 frequency points. The GPDD and DDD construction time was much less than the ac evaluation time among the total runtime of less than 2 seconds. The column "Mem" lists the memory consumptions for running the two benchmarks.

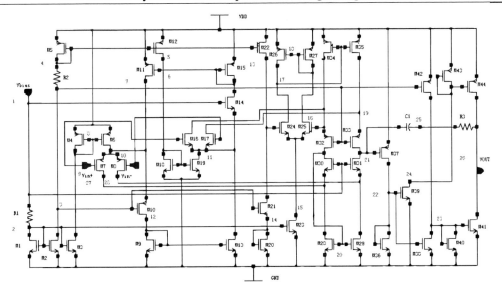

Figure 11. Op-amp 3: A MOS op-amp containing 44 transistors [30].

Table 1. Performance of the symbolic stamp method

Op-amp	#FET (T)	#Symb (GPDD)	#Symb (DDD)	Matrix Size	$|GPDD|$	$|DDD|$	Time (sec)	Mem (MB)
Op-amp 2	24	12	104	18×18	481	70,129	1.81	70
Op-amp 3	44	12	140	28×28	481	45,716	1.50	91

In the evaluation phase, the small-signal device parameter values corresponding to each MOSFET are passed to the multi-root GPDD and evaluated to get the stamp values specifi to a MOS device. The hierarchical treatment results in a 18×18 MNA matrix containing 104 nonzeros (symbols) for *op-amp 2* and a 28×28 MNA matrix containing 140 nonzeros (symbols) for *op-amp 3*. The DDD routine created a DDD of 70,129 vertices for the MNA matrix of op-amp 2 and a DDD of 45,716 vertices for the MNA matrix of op-amp 3. Note that the DDD size $|DDD|$ (i.e., the total number of DDD vertices) depends on both the matrix size and the specifi symbol order. A larger matrix size with a good symbol order could possibly result in a smaller DDD size, as seen in Table 1. We note that neither of the non-hierarchical implementations of DDD and GPDD could solve the two op-amp circuits exactly in our experiment.

A number of hierarchical symbolic analysis methods have been published in the literature. Table 2 summarizes a comparison of the performances based on the published data. Except for the work [31] which implemented an *approximate* hierarchical analysis scheme, the symbolic stamp approach of [27] was able to solve the largest op-amp circuits (containing more than 40 MOS transistors) *exactly*.

Table 2. Performance comparison of representative hierarchical methods

Publication	Max Ckt Size (#Q/#J/#M)	Method	Accuracy
[21]	20 (BJT)	DDD + Schur decomp.	Exact
[32]	26 (BJT)	Regularity + Sharing	Exact
[31]	83 (MOS)	Ckt reduction + Two-graph	Approx.
[24]	26 (BJT)	DDD + De-cancellation	Exact
[27]	44 (MOS)	DDD + GPDD	Exact

4. Symbolic AC Sensitivity Analysis

In circuit design, the designer is interested in the variational dependence of a design metric with respect to (w.r.t.) one or many selected design parameters. Such variational dependence can be characterized by sensitivity analysis.

There are several notions of sensitivity commonly used in circuit design. The popular SPICE software can compute the dc or ac sensitivity of a selected variable with respect to all circuit parameters. The computation is based on a technique called *adjoint network analysis* [33].

Circuit sensitivity analysis has been a very basic concept familiar to most circuit designers. However, little agreement has been reached on what kind of circuit sensitivity should be provided by a design tool and how to use the variety of possible sensitivities. A more useful but computationally harder sensitivity is the *pole/zero sensitivity*, which was discussed by [34] and [35]. Lee et al. in [34] addressed a moment matching method for faster but approximate computation of the pole/zero sensitivity, while the work by Huang et al. [35] addressed the application of pole/zero sensitivity for nonlinear behavioral model construction.

It is well-known that solving all poles and zeros of an analog circuit requires a good amount of computation. Finding the sensitivity of poles and zeros with respect to one or multiple circuit parameters would require more laborious computation. The computation obstacle motivated Lee *et al.* to apply the moment matching technique [36] to approximate computation of pole/zero sensitivities. It was demonstrated in [34] that even approximately computed pole/zero sensitivities showed good correlation with the real circuit behavior, hence could be used for the purpose of design optimization.

The nonlinear behavioral model construction is a more involved issue as discussed in [35], because one has to change the operating conditions to characterize the nonlinearity. Prominent nonlinearities existing in the circuit are identifie by the root localization and signal-path tracing techniques proposed in that work, for which the root sensitivities played an important role. Again, the pole/zero sensitivity is the most informative in revealing the critical dependence of circuit behavior on circuit elements and their connections.

In this section, the author would like to demonstrate the effectiveness of an alternative approach by using the symbolically computed ac sensitivity to characterize the circuit behavior. By *symbolic ac sensitivity* we mean the sensitivity of a symbolic transfer function with respect to any circuit parameter, including MOS device sizes. It will be shown that symbolic ac sensitivity can actually exhibit pole/zero locations and their dependence on the device parameters, which is essentially a new characterization of pole/zero sensitivity. But the computation cost is much lower and the intuitiveness is much better, as we shall see later in this section.

4.1. Sensitivity Identities

Symbolic ac sensitivity takes an analytical approach, hence some analytical identities valid for normalized sensitivity can help provide analytical justif cation for those graphically exposed features when the symbolic ac sensitivities are plotted. The well-known *normalized sensitivity* is def ned by

$$\text{Sens}(y, x) := \frac{x\,dy}{y\,dx} = \frac{d\ln(y)}{d\ln(x)}, \tag{15}$$

where the argument of the logarithmic function could be complex numbers.

Listed below are some very basic identities valid for the normalized sensitivity. They are useful for sensitivity computation in the symbolic setting. Because the normalized sensitivity is def ned in terms of the "$\ln(\cdot)$" function, the sensitivities on the operations of *multiplication, division*, or *natural exponential function* have simple expressions. The proof of some identities can be found in [37].

Identity 1. *Scaling $H(s)$ by a constant does not change the sensitivity. That is,*

$$\text{Sens}(cH(s), v) = \text{Sens}(c, v) + \text{Sens}(H(s), v) = \text{Sens}(H(s), v), \tag{16}$$

where c is a constant.

Identity 2. *Let $G = 1/R$. Then*

$$\text{Sens}(H(s), R) = -\text{Sens}(H(s), G). \tag{17}$$

Identity 3. *Let $H(s) = 1/X(s)$. Then*

$$\text{Sens}(H(s), p) = -\text{Sens}(X(s), p). \tag{18}$$

Identities 2 and 3 lead to

Identity 4. *Let $H(s) = 1/X(s)$ and $G = 1/R$. Then*

$$\text{Sens}(H, R) = \text{Sens}(X, G). \tag{19}$$

Identity 5. *Let $H(s) = N(s)/D(s)$*

$$\text{Sens}(H(s), p) = \text{Sens}(N(s), p) - \text{Sens}(D(s), p). \tag{20}$$

Viewing 's' as a constant, identity 1 yields

Identity 6.

$$\text{Sens}(H(s), C) = \text{Sens}(H(s), Cs). \tag{21}$$

24 Guoyong Shi

The sensitivities of the magnitude and phase of $H(s)$ are given by the following two identities, which are very useful in design optimization.

Identity 7. *The sensitivity of the magnitude of $H(s)$ is*

$$\text{Sens}(|H(s)|, v) = \text{Re}\left\{\text{Sens}(H(s), v)\right\}. \tag{22}$$

Identity 8. *The sensitivity of the phase of $H(s)$ is*

$$\text{Sens}(\angle H(s), v) = \frac{1}{\angle H(s)} \text{Im}\left\{\text{Sens}(H(s), v)\right\}. \tag{23}$$

Proof. We obtain from $H(s) = |H(s)|e^{j\angle H(s)}$ that

$$
\begin{aligned}
\text{Sens}(H(s), v) &= \frac{\partial \ln\left(|H(s)|e^{j\angle H(s)}\right)}{\partial \ln v} \\
&= \frac{\partial \ln|H(s)|}{\partial \ln v} + j\frac{\partial \angle H(s)}{\partial \ln v} \\
&= \text{Sens}(|H(s)|, v) + j\angle H(s)\,\text{Sens}(\angle H(s), v).
\end{aligned}
$$

Hence, identities 7 and 8 follow. $\square$

Remark 1. *From the proof we see that the imaginary part of $\text{Sens}(H(s), v)$ is $\angle H(s)\,\text{Sens}(\angle H(s), v)$, which is the normalized phase sensitivity multiplied by the phase. We may simply use the imaginary part of $\text{Sens}(H(s), v)$ for the phase sensitivity which is just a modification in normalization.*

Remark 2. *Suppose we would like to have the derivative of a circuit parameter v w.r.t. the phase $\angle H(s)$ and run an optimization on v to improve the phase margin. The derivative $\partial v/\partial(\angle H(s))$ can simply be obtained from $(v/\text{Im}\left\{\text{Sens}(H(s), v)\right\})$, as can be justified easily.*

Remark 3. *Some of the identities listed above are intended for use in symbolic analysis. For example, C is the symbol for a capacitor in the GPDD representation. At the time of ac evaluation, this symbol should be substituted by Cs, which also appears in the expanded product terms. But for sensitivity analysis, one may simply operate on the GPDD vertex whose value is given by C. Similar interpretation applies to the identities 2, 3, and 4, etc. (see [37]). The awareness of this observation simplifies implementation a lot.*

4.2. Symbolic Sensitivity of Transfer Functions

Suppose a rational transfer function is written in the following *gain-normalized* form

$$H(s) = G\frac{\prod_{i=1}^{m}(\zeta_i s + 1)}{\prod_{j=1}^{n}(\tau_j s + 1)}, \tag{24}$$

where $G = H(0)$ is the dc gain and both ζ_i and τ_j are complex numbers in general. We let $z_i = -1/\zeta_i$ denote a zero and $p_i = -1/\tau_i$ a pole.

Let v be the sensitivity parameter of interest. Clearly, the sensitivity of $H(s)$ w.r.t. the parameter v involves the sensitivities of a generic term $(as+1)$ w.r.t. v. Therefore, it would be helpful to take a look at $\text{Sens}(as+1,v)$ first before we proceed.

Suppose the coefficient a depends on v. Then the sensitivity of the factor $(as+1)$ w.r.t. v is given by

Identity 9. *Let $p = -1/a$. Then*

$$\text{Sens}(as+1, v) = \text{Sens}(a, v)\left(\frac{as}{as+1}\right) = -\text{Sens}(p, v)\left(\frac{s}{s-p}\right). \tag{25}$$

Proof.

$$\begin{aligned}
\text{Sens}(as+1, v) &= \left(\frac{\partial a}{\partial v}\right)\left(\frac{\partial (as+1)}{\partial a}\right)\left(\frac{v}{as+1}\right) \\
&= \left(\frac{\partial a}{\partial v}\right)\left(\frac{vs}{as+1}\right) = \left(\frac{\partial a}{\partial v} \cdot \frac{v}{a}\right) \cdot \left(\frac{as}{as+1}\right) \\
&= \text{Sens}(a, v)\left(\frac{as}{as+1}\right)
\end{aligned}$$

The second equality follows from the sensitivity identities. $\qquad\square$

Remark 4. *It is interesting to note that the sensitivity of a pole/zero term $(as+1)$ can be expressed by the sensitivity of the pole/zero $\text{Sens}(a, v) = -\text{Sens}(p, v)$ multiplied by a linear fractional function $as/(as+1)$. It is known in Complex Analysis that a linear fractional transformation is a conformal mapping that maps the $j\omega$ axis to a circle. This property will be used later in visual pole/zero characterization in sensitivity plots.*

The preceding sensitivity identity provides a linear combination basis function for expressing the sensitivity of a general network function $H(s)$, which is stated by the next identity.

Identity 10. *The sensitivity of $H(s)$ with respect to a parameter v is expressed by*

$$\begin{aligned}
\text{Sens}(H(s), v) = \text{Sens}(G, v) + \sum_{i=1}^{m} \text{Sens}(\zeta_i, v)\left(\frac{\zeta_i s}{\zeta_i s + 1}\right) \\
- \sum_{j=1}^{n} \text{Sens}(\tau_j, v)\left(\frac{\tau_j s}{\tau_j s + 1}\right). \tag{26}
\end{aligned}$$

Proof. Applying the basic sensitivity identities to the factorized transfer function (24), we get

$$\begin{aligned}
\text{Sens}(H(s), v) = \text{Sens}(G, v) + \sum_{i=1}^{m} \text{Sens}(\zeta_i s + 1, v) \\
- \sum_{j=1}^{n} \text{Sens}(\tau_j s + 1, v). \tag{27}
\end{aligned}$$

Then an application of the monomial sensitivity identity 9 leads to the identity 10. $\qquad\square$

The importance of the sensitivity formula (26) for a general transfer function is explained next. The expression is in the form of a linear combination of linear fractional transformations $\zeta_i s/(\zeta_i s+1)$ or $\tau_j s/(\tau_j s+1)$, as the basis functions, multiplied by the zero/pole sensitivities $\text{Sens}(\zeta_i, v)$ and $\text{Sens}(\tau_j, v)$, as the coefficients. Although the pole/zero sensitivities are not explicitly calculated in the expression, their appearance as coefficients can roughly be visualized in the plots of sensitivity, as we shall see later. This interesting connection between the sensitivity of a transfer function and the pole/zero sensitivity has never been studied in the literature.

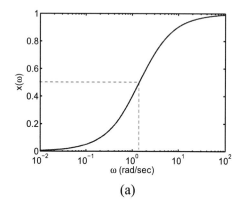
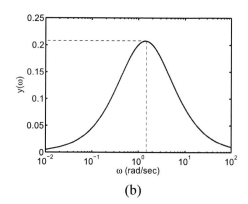

(a)　　　　　　　　　　　　　　　　　　　(b)

Figure 12. For $p = -1 - j$ (i.e., $\alpha = 1$ and $\beta = -1$), $x(\omega)$ is plotted in (a) and $y(\omega)$ is plotted in (b).

The trajectory of the linear fractional basis function $h(s,a) := as/(as+1)$ for $s = j\omega$, $\omega \in (-\infty, +\infty)$, is a circle in the complex plane. Assuming $p = -1/a = \alpha + j\beta$ be a pole or zero with $\alpha > 0$, we obtain that

$$h_\beta(j\omega) = \frac{j\omega}{j\omega + \alpha - j\beta} = x_\beta(\omega) + jy_\beta(\omega), \tag{28}$$

where $x_\beta(\omega)$ and $y_\beta(\omega)$ are respectively the real and imaginary parts of $h_\beta(j\omega)$. Since the complex poles or zeros normally appear in conjugate pairs (i.e., $p = -\alpha \pm j\beta$), the dependence on β is written explicitly.

For an example, the real and imaginary part plots of $h_\beta(j\omega)$ for $p = -1 - j$ are shown in Fig. 12. We see that the plots of $x_\beta(\omega)$ and $y_\beta(\omega)$ exhibit a kind of *mode* in the selected frequency range. Here by *mode* we refer to the mid-point of the transition ramp of $x_\beta(\omega)$ or the peak of $y_\beta(\omega)$. For this example, the *mode* appears exactly at ω equal to the radius of the pole (or zero), i.e., $\omega = |p| = \sqrt{\alpha^2 + \beta^2} = \sqrt{2}$. In case of a real p, i.e., $\beta = 0$, the mode will be seen right at the location of $\omega = |p| = |\alpha|$.

Note that the complex number $h_\beta(j\omega)$ is scaled by a coefficient of $\text{Sens}(p, v)$ in the formula (26). Hence, the curves of $x_\beta(\omega)$ and $y_\beta(\omega)$ scaled by the same coefficient $\text{Sens}(p, v)$ contribute in linear combination to the whole sensitivity function $\text{Sens}(H(s), v)$ as a complex number. The sensitivity coefficient $\text{Sens}(p, v)$ is complex if p is. A complex sensitivity coefficient would possibly distort the modes in the real and imaginary parts of $x_\beta(\omega)$ and $y_\beta(\omega)$, but not drastically. We expect that by plotting the real and imaginary parts of

Sens$(H(j\omega), v)$ versus the frequency ω and inspect the *modes* exhibiting in the curves, it is possible to tell roughly the locations of dominant poles and zeros and even the magnitude of sensitivity the parameter v affects the poles and zeros. For the case of real poles or zeros, their locations are exactly where the modes are if the nearby modes do not interfere each other.

We would like to leave analytical details for above observation to other publications. Instead, we go through some simple circuit examples to illustrate the interesting sensitivity *mode* phenomenon.

4.3. Simple Circuit Examples

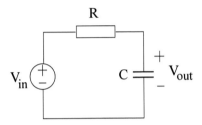

Figure 13. One-stage RC circuit.

The voltage transfer function of the RC circuit given in Fig. 13 from V_{in} to V_{out} is

$$H_1(s) = \frac{1}{RCs+1} = \frac{1}{\tau s+1}, \qquad (29)$$

where $\tau = RC$ is the time constant. $H_1(s)$ has a unity dc gain and a real pole at $p = -1/\tau$. The sensitivity of $H_1(s)$ w.r.t. the resistor R is

$$\text{Sens}(H_1(s), R) = -\frac{RCs}{RCs+1} = -\frac{\tau s}{\tau s+1}. \qquad (30)$$

The real and imaginary parts of Sens$(H_1(j\omega), R)$ are plotted in Fig. 14 for $\tau = 2$. We see that $x(\omega) := \text{Re}\{\text{Sens}(H_1(j\omega), R)\} = \text{Sens}(|H_1(j\omega)|, R)$ reaches the mid-value $x(\omega) = 0.5$ at $\omega = 1/\tau = 0.5$ rad/sec and $y(\omega) = \text{Im}\{\text{Sens}(H_1(j\omega), R)\}$ reaches the negative peak at $\omega = 0.5$ rad/sec as well. Therefore, the sensitivity mode coincides with the location of the real pole.

According to the general sensitivity formula in equation (10), the sensitivity coefficient for the pole $p = -\tau^{-1}$ w.r.t. the resistor R is Sens$(\tau, R) = 1$, meaning that increasing the value of R would increase the time constant $\tau = RC$ and equivalently reduces the pole. This is seen from the plot of Sens$(|H_1(j\omega)|, R)$ shown in Fig. 14(a), where the magnitude sensitivity is negative, meaning that increasing R would decrease the magnitude of $|H_1(j\omega)|$, but the decreasing rate varies over the frequency, lower nearby the dc band while higher toward the high frequency. The effect is that the pole (or bandwidth) is reduced. The phase sensitivity curve also can be interpreted analogously; the phase of 0 degree at the lower frequency and the phase of -90 degrees at the high frequency are not sensitive to the change of R, but the phase transition (from 0 to -90 degrees) nearby the pole is more sensitive, the phase is shifted lower, corresponding to reducing the pole as R is increased.

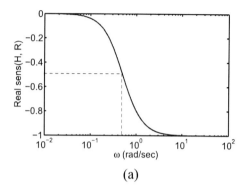
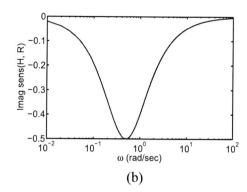

Figure 14. The real and imaginary parts of $\text{Sens}(H_1(j\omega), R)$ are shown in (a) and (b) respectively for the circuit in Fig. 13 with $\tau = RC = 2$.

The above observation reveals that the sensitivity curves plotted in the frequency domain exhibit not only the position of a real pole, but also the trend and strength of the frequency response variation caused by changing the element value. Such variational information are believed to be valuable to analog designers.

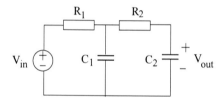

Figure 15. Two-stage RC circuit.

For the second example, the voltage transfer function of the two-stage RC circuit shown in Fig. 15 from V_{in} to V_{out} is given by

$$H_2(s) = \frac{G_1 G_2}{C_1 C_2 s^2 + (C_1 G_2 + C_2 G_1 + C_2 G_2)s + G_1 G_2}, \quad (31)$$

which has a unity dc gain again and two real poles. The sensitivity of $H_2(s)$ w.r.t. C_1 is calculated as

$$\text{Sens}(H_2(s), C_1) = -\frac{C_1(C_2 s^2 + G_2 s)}{C_1 C_2 s^2 + (C_1 G_2 + C_2 G_1 + C_2 G_2)s + G_1 G_2}. \quad (32)$$

The real and imaginary parts of $\text{Sens}(H_2(j\omega), C_1)$ are plotted in Fig. 16 for $C_1 = C_2 = G_1 = 1$ and $G_2 = 2$. The positions of two real poles can be seen roughly from the two downward slopes of $\text{Re}\{\text{Sens}(H_2(j\omega), C_1)\}$ or the two modes of $\text{Im}\{\text{Sens}(H_2(j\omega), C_1)\}$. The two real poles are computed to be $p_1 = -0.4384$ rad/sec and $p_2 = -4.5616$ rad/sec, which roughly coincide with the mode locations.

Although it is a little hard to discern the exact locations of two closely placed poles as in this example, it does not prevent us from noticing the graphical exposition of zeros/poles in the sensitivity plots, and even observing the relative strength of the sensitivities. In case the exact poles/zeros are desired, they can be obtained by running a commercial SPICE simulator.

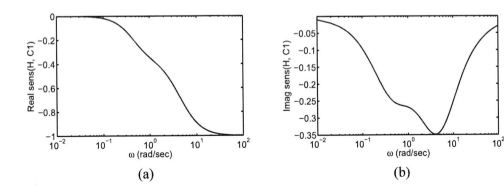

Figure 16. The real and imaginary parts of $\text{Sens}(H_2(j\omega), C_1)$ are shown in (a) and (b) respectively for the circuit in Fig. 15 with $C_1 = C_2 = G_1 = 1$ and $G_2 = 2$.

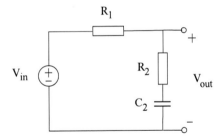

Figure 17. An RC circuit with one pole and one zero, where $R_1 = R_2 = 1\Omega$ and $C_2 = 1F$.

The third example given in Fig. 17 has a real zero and a real pole. The voltage transfer function from V_{in} to V_{out} is given by

$$H_3(s) = \frac{R_2 C_2 s + 1}{(R_1 + R_2) C_2 s + 1}. \tag{33}$$

This transfer function has a unity dc gain, one real zero at $z = -1/(R_2 C_2)$, and one real pole at $p = -1/((R_1 + R_2)C_2)$. Both the zero and pole depend on C_2, which is seen clearly in the sensitivity plots of $\text{Sens}(H_3(s), C_2)$ shown in Fig. 18 for the chosen parameters of $R_1 = R_2 = 1\Omega$ and $C_2 = 1F$. Despite the proximity of the pole and zero, the corresponding modes are clearly seen in the plots. It is not important whether or not the exact pole or zero can be located. What is important is the disclosure of how the element C_2 affects the existing pole and zero as seen from sensitivity curves, while such information is not available by plotting only the frequency response curves. For this example, the sensitivity curves suggest us to increase the value of C_2 if we would like to shift both pole and zero toward the lower frequency because the sensitivity of the magnitude $|H(s)|$ w.r.t. C_2 is negative.

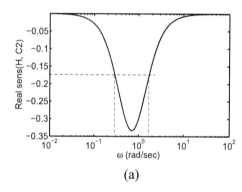 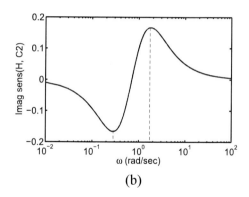

(a)   (b)

Figure 18. The real and imaginary parts of $\text{Sens}(H_3(j\omega), C_2)$ are shown in (a) and (b) respectively for the circuit in Fig. 17 with $R_1 = R_2 = 1\Omega$ and $C_2 = 1F$.

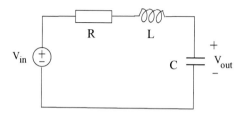

Figure 19. RCL circuit with a pair of conjugate poles.

To see the effect of conjugate poles, the fourth example shown in Fig. 19 is examined. The voltage transfer function of the circuit from V_{in} to V_{out} is given by

$$H_4(s) = \frac{1}{LCs^2 + RCs + 1}. \tag{34}$$

This transfer function has a unity dc gain and two conjugate poles for the element values satisfying $R^2 < 4L/C$. The real and imaginary parts of $\text{Sens}(H_4(j\omega), L)$ are plotted in Fig. 20 for the parameters $R = 2\Omega$, $L = 2H$, and $C = 1F$. The time constants are $a_{1,2} = 1 \pm j$, with the radius of pole equal to $|p| = 0.707$. The sensitivity plots exhibit warped modes around this location, i.e., a *slope* in the real part and a *valley* in the imaginary part, but the location is not exactly at $|p| = 0.707$. The reason is that, in view of the equation (10), the sensitivity coeffcient $\text{Sens}(a, v)$ is now complex, it thus modifes the complex linear fraction $a(j\omega)/(a(j\omega)+1)$ which is a circle in the complex plane, causing the mode dislocation in the plots of real and imaginary parts. A lengthy analysis in this respect will be presented elsewhere.

However, even the existence of such a location shift in pole for the case of conjugate poles does not affect how we utilize the sensitivity information on the pole. The plots tell us confrmatively that increasing L would move the conjugate poles toward lower frequency. Also, a positive hump in the real part plot tells us that increasing L would increase the amplitude response at that frequency, which is a fact that matches the circuit property: increasing L reduces the damping factor, thus increasing oscillation around the resonance frequency.

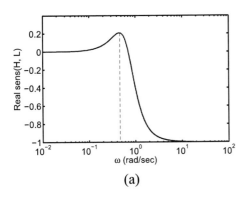 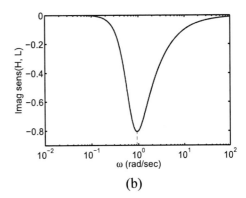

Figure 20. The real and imaginary parts of $\mathrm{Sens}(H_4(j\omega),L)$ are shown in (a) and (b) respectively for the circuit in Fig. 19 with $R = 2\Omega$, $L = 2H$, and $C = 1F$.

5. Symbolic Sensitivity Computation and Application

While the symbolic sensitivities for those small circuits can be calculated manually without diff culty, deriving symbolic sensitivities for an op-amp circuit containing many semiconductor devices is a nontrivial task. If the designer does not have a symbolic tool at hand, he/she may consider to use a SPICE simulator by running ac analysis twice for a parameter taking two close values, then calculate the f nite difference and apply normalization to get the numerical ac sensitivity. Plotting the real and imaginary parts versus the frequency ω, one may get the approximate sensitivity curves. Disregarding the numerical errors that might appear in the difference calculation, the sensitivity curves show similar shapes as we have seen in the discussion on the simple circuits.

The advantage of symbolic sensitivity, if available, lies in the fact that one can obtain the sensitivities with respect to many parameters provided that a symbolic transfer function exists. By the BDD-formulated exact symbolic methods we reviewed earlier, deriving the symbolic ac sensitivity would not be diff cult anymore. Moreover, by applying the hierarchical analysis strategy using symbolic stamps (introduced earlier in Section 3.), large op-amp circuits with more than 40 MOS transistors can be analyzed exactly for symbolic ac sensitivity. In this section, we demonstrate by an example how the ac sensitivity curves would look like for op-amp circuits and how to use such sensitivity measures for circuit optimization. More design oriented actions can be proposed in the perspective of a design environment supported by an interactive graphical user interface (GUI). But these are not the subjects to be discussed in this chapter.

The symbolic ac sensitivity computation was mentioned simply in the paper [16], but without a detailed expansion on that. A preliminary study on sizing by symbolic sensitivity was presented in the work of Yang et al. [38,39], where a variant of DDD was developed. In both [38,39] the sensitivity was used in the gradient search algorithm for optimizing a sizing object function, not in the form of sensitivity plots that we have been discussing. The work of [37] made an investigation on the symbolic sensitivity to the small-signal parameters using GPDD by plotting the sensitivity curves, but did not get into much details on the

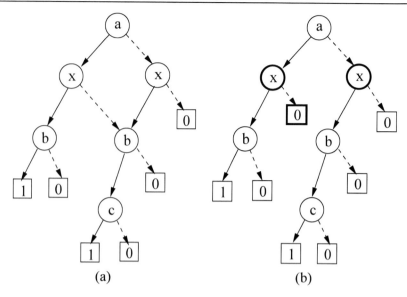

Figure 21. (a) The BDD before sensitivity operation. (b) The BDD after the sensitivity operation.

pole/zero modes and their connection to circuit optimization. The very recent work [40] further investigated the MOSFET sizing sensitivity by visualizing the sensitivity plots. It is discovered for the first time that the dependence of poles/zeros on the MOS devices can be disclosed explicitly in the sensitivity plots as well. That research has in a sense established a fundamental tool mechanism for such design automation tasks as semiconductor device sizing, which has always been considered hard for automation. More research effort in this direction is believed to be worthwhile for a full realization of analog IC design automation.

The implementation of sensitivity calculation in a BDD can be unexpectedly easy, especially in a GPDD, which can be illustrated by a simple example. Let $E(a,x,b,c) = axb + xbc + abc$ be a sum-of-product expression, where the variables are ordered in $a < x < b < c$. It is readily derived that

$$\text{Sens}(E,x) = \frac{x}{E} \cdot \frac{\partial E}{\partial x} = \frac{x}{E}(ab+bc) = \frac{axb+xbc}{E} = \frac{E - E|_{x=0}}{E}, \quad (35)$$

where in the last expression the subtraction of $E|_{x=0}$ means eliminating those product terms not involving x. Hence, the normalized sensitivity of the SOP expression w.r.t. the variable x can be calculated by retaining those product terms involving x divided by the SOP itself.

The sensitivity calculation for an SOP represented by a BDD can be implemented easily with only slight data structure modifications. This fact hold true in particular for the GPDD representation of a symbolic transfer function. Let us use an example to illustrate the simple operations applied on a BDD. In Fig. 21, the left BDD represents the SOP expression of $E(a,x,b,c) = axb + xbc + abc$ and the right BDD represents the BDD after the sensitivity operation w.r.t. x is performed (not divided by E yet). What is done in the BDD is just to terminate the "dashed" arrows of all vertices named "x" to vertex "0" (those already terminating at "0" are retained.)

The implementation of the sensitivity operation on a DDD is analogous, but requires identifying more DDD vertices that are dependent on the sensitive variable, because the DDD vertex variables are the MNA matrix elements, not directly the circuit parameters. More details on implementation are discussed in the recent work [41].

The ac sensitivity w.r.t. the semiconductor device sizes can be derived analytically provided that the small-signal parameters in terms of the device sizes are available explicitly. One might have to divide the device operation region into several parts so that the approximate analytical expressions in terms of the device dimension can be derived. By assuming that all MOS devices are sized in their saturation region, the work [40] analyzed some op-amp sensitivities with respect to device sizes.

The two stage op-amp with RC compensation shown in Fig. 1 is now used to demonstrate the use of ac sensitivity for device size optimization. The op-amp circuit consists of eight MOSFETs. Initially, the transistors are sized quite arbitrarily to be biased in the saturation region without considering the design goal too much.

The symbolic simulator developed in the author's laboratory can calculate the ac sensitivities with respect to all MOS device sizes. The real and imaginary parts of the ac sensitivity to all device sizes are plotted in Figs. 22, where the "modes" (either slopes or peaks) indicate the possible pole/zero locations. It is interesting to observe that some sensitivity curves like those with $M6$ and Cc are very oscillatory in the visible frequency range. That would indicate clearly that the existing dominant poles and zeros are more sensitive to the output device $M6$ and the compensation capacitor Cc, which means that sizing these two devices can change the placement of the relevant poles and zeros most effectively. On the other hand, sizing MOSFETs $M1$-$M2$ (as a pair) and $M6$ can most effectively increase the dc gain.

Besides providing the intuitive indication of dependence of circuit behavior on devices, the ac sensitivities can also be used for the purpose of specif c optimization, such as adjusting a selected device for improving the phase margin. For example, we can choose to optimize the nulling resistor value Rz in the op-amp 1 (Fig. 1) to achieve a preset phase margin.

After the initial sizing, the phase margin of the amplif er had a phase margin about 25 degree by HSPICE simulation. If we plot the dominant poles and zeros (see Fig. 23(a)), we see that the f rst zero is apart from the second dominant pole. In this case, adjusting the value of the nulling resistor Rz can possibly improve the phase margin. Our goal is to increase the phase margin above 60 degrees (a typical design requirement). The update formula of Rz is given by

$$
\begin{aligned}
R_z^{(n+1)} &= R_z^{(n)} + \frac{\partial(R_z^{(n)})}{\partial(\angle H(s))}\Delta(\angle H(s)) \\
&= R_z^{(n)} + \frac{R_z^{(n)} \cdot \Delta(\angle H(s))}{\mathrm{Im}\left\{\mathrm{Sens}(H(s), R_z^{(n)})\right\}},
\end{aligned}
\tag{36}
$$

where the sensitivity $\mathrm{Sens}(H(s), R_z^{(n)})$ is always computed at the unity-gain frequency. (See Remark 2 in subsection 4.1. for an interpretation of the iteration formula.)

After four iterations (see Table 3), the phase margin reaches 60 degrees. Plotting the poles and zeros again in Fig. 23(b) we see that the f rst zero now has been moved closer to

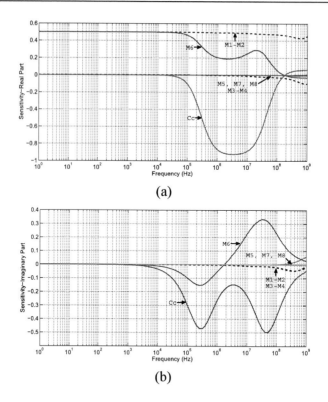

Figure 22. (a) Plot of $\mathrm{Re}\{\mathrm{Sens}(|H(s)|, W_k)\}$. (b) Plot of $\mathrm{Im}\{\mathrm{Sens}(\angle H(s), W_k)\}$.

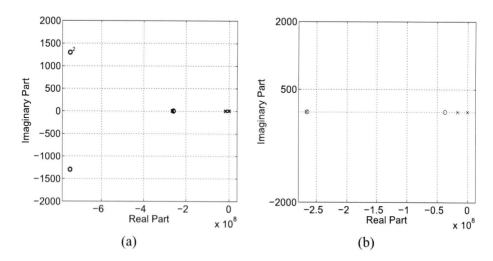

Figure 23. Dominant poles and zeros by HSPICE: (a) Pole and zero locations before R_z optimization. (b) Pole and zero location after R_z optimization.

the second dominant pole after Rz optimization. A similar optimization procedure using the approximately computed pole-zero sensitivity by AWE was proposed in [34]. Comparing to that work, the amount of computation required for device optimization by the symbolic

Table 3. Sensitivity-based resistor iteration

Iteration	$R_z(k\Omega)$	$\angle H(s)(deg)$	$Sens(H(s), R_z)$	$\Delta R_z\ (\Omega)$
1	1.00	−155.6	10.2852	3,460
2	4.46	−125.0	30.3875	739
3	5.20	−120.5	28.2814	95.5
4	5.29	−120.0		

ac sensitivity is much less. More importantly, such optimization procedure requires the repeated computation of sensitivity, for which a symbolic tool is certainly more efficient

Finally, we mention that the state-of-the-art symbolic tool for exact circuit analysis has reached an advanced level of analyzing CMOS circuits containing 40 to 50 semiconductor devices, sufficien for most practical needs. The recent work [41] reported that the symbolic ac sensitivity could be carried out for the two large op-amp circuits as shown in Fig. 10 (op-amp 2 containing 24 MOSFETs) and in Fig. 11 (op-amp 3 containing 44 MOSFETs) by the hierarchical analysis method based on symbolic stamps. The sensitivity plots of the op-amp 2 are shown in Fig. 24 for some selected devices, from which we can see clearly that some devices are more sensitively related to the dominant poles and zeros than others.

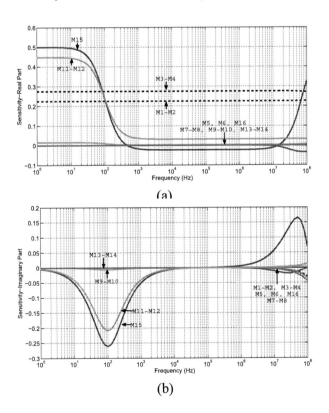

Figure 24. (a) The real parts of ac sensitivities; (b) The imaginary parts of ac sensitivities.

With the advanced data structure such as two layers of BDDs being used in the new generation of exact symbolic analysis, the memory and time consumption for symbolic network function construction, sensitivity computation, and numerical evaluation can be managed at the lowest level. As an indicative measure, the runtime performance in the listed time splits is collected in Table 4 for the three op-amps circuits studied in this chapter. The numerical evaluation time was measured over running the ac analysis and sensitivity analysis with 100 frequency points. Even for the largest op-amp 3 containing 44 MOS transistors, the time and memory cost is not overwhelming at all. The computer used for experiment was again a desktop with AMD Athlon64 2.20GHz processor and 2GB memory. It indicates that the state-of-the-art symbolic technology is almost ready for industrial use.

Table 4. Performance of hierarchical sensitivity analysis

Circuit	Op-amp 1	Op-amp 2	Op-amp 3
No. transistors	8	24	44
GPDD build time (sec.)	0.1	0.1	0.1
DDD build time (sec.)	0.01	0.18	0.3
Evaluation time (sec.)	0.28	23.23	57.02
Total time (sec.)	0.38	23.51	57.42
Memory (MB)	56	101	153

6. Conclusion

This chapter has presented a very detailed exposition on the new developments of symbolic circuit analysis methods in the past decade. The binary decision diagram (BDD) based techniques are emphasized, because they represent the most advanced computation technology that can handle enumeration-based large NP-complete problems. With the innovations achieved recently in the area of symbolic circuit analysis, those large op-amp circuits once unsolvable exactly by a symbolic tool can now be solved with efficiency. Along with the survey of the newly developed analysis methods, the new notion of symbolic ac sensitivity is introduced as well. The design oriented aspects are emphasized, including the underlying relation between ac sensitivity and pole/zero sensitivity, the utilization of information revealed by the sensitivity plots, and the use of repeatedly computed sensitivity for iterative parameter optimization. It is important to be aware that the ac sensitivity analysis must be based on an *exact* symbolic transfer function. It is expected that the new advancements achieved in the past decade on symbolic circuit analysis can positively contribute to the research of analog design automation, which is still at a very low level after many decades of research. In the author's perspective, no computational technology other than the symbolic approach can provide the most fundamental solution to analog design automation.

References

[1] P. R. Gray and R. Meyer, "MOS operational amplifer design – a tutorial overview," *IEEE Journal of Solid-State Circuits*, vol. SC-17, no. 6, pp. 969–982, Dec. 1982.

[2] K. P. Ho, C. F. Chan, C. S. Choy, and K. P. Pun, "Reversed nested Miller compensation with voltage buffer and nulling resistor," *IEEE Journal of Solid-State Circuits*, vol. SC-38, no. 10, pp. 1735–1738, Oct. 2003.

[3] R. E. Bryant, "Graph-based algorithms for boolean function manipulation," *IEEE Trans. on Computers*, vol. C-35, no. 8, pp. 677–691, 1986.

[4] G. Gielen and W. Sansen, *Symbolic Analysis for Automated Design of Analog Integrated Circuits*. Norwell, MA: Kluwer Academic Publishers, 1991.

[5] G. Nebel, U. Kleine, and H. J. Pfeiderer, "Symbolic pole/zero calculation using SANTAFE," *IEEE Journal of Solid-State Circuits*, vol. 30, no. 7, pp. 752–761, July 1995.

[6] O. Guerra, J. D. Rodríguez-García, F. V. Fernández, and A. Rodríguez-Vázquez, "A symbolic pole/zero extraction methodology based on analysis of circuit time-constants," *Analog Integrated Circuits and Signal Processing*, vol. 31, pp. 101–118, 2002.

[7] C. W. Ho, A. E. Ruehli, and P. A. Brennan, "The modifed nodal approach to network analysis," *IEEE Trans. on Circuits and Systems*, vol. CAS-22, no. 6, pp. 504–509, 1975.

[8] G. Shi, W. Chen, and C. J. R. Shi, "A graph reduction approach to symbolic circuit analysis," in *Proc. Asia South-Pacific Design Automation Conference (ASPDAC)*, Yokohama, Japan, Jan. 2007, pp. 197–202.

[9] Q. Yu and C. Sechen, "A unifed approach to the approximate symbolic analysis of large analog integrated circuits," *IEEE Trans. on Circuits and Systems – I: Fundamental Theory and Applications*, vol. 43, no. 8, pp. 656–669, 1996.

[10] W. Chen and G. Shi, "Implementation of a symbolic circuit simulator for topological network analysis," in *Proc. Asia Pacific Conference on Circuits and Systems (APC-CAS)*, Singapore, Dec. 2006, pp. 1327–1331.

[11] P. M. Lin, *Symbolic Network Analysis*. New York: Elsevier, 1991.

[12] G. Gielen, P. Wambacq, and W. M. Sansen, "Symbolic analysis methods and applications for analog circuits: A tutorial overview," *Proceedings of the IEEE*, vol. 82, no. 2, pp. 287–303, February 1994.

[13] R. E. Bryant and J. H. Kukula, "Formal methods for functional verifcation," in *The Best of ICCAD, 20 Years of Excellence in Computer-Aided Design*, A. Kuehlmann, Ed. Norwell, MA, USA: Kluwer Academic Publishers, 2003, pp. 3–15.

[14] K. S. Brace, R. L. Rudell, and R. E. Bryant, "Eff cient implementation of a BDD package," in *Proc. 27th ACM/IEEE Design Automation Conference*, Orlando, FL, 1990, pp. 40–45.

[15] S. Minato, *Binary Decision Diagrams and Applications for VLSI CAD.* Norwell, MA: Kluwer Academic, 1996.

[16] C. J. R. Shi and X. D. Tan, "Canonical symbolic analysis of large analog circuits with determinant decision diagrams," *IEEE Trans. on Computer-Aided Design of Integrated Circuits and Systems*, vol. 19, no. 1, pp. 1–18, January 2000.

[17] S. X. D. Tan, "Symbolic analysis of analog integrated circuits by boolean logic operations," *IEEE Trans. on Circuits and Systems - II: Express Briefs*, vol. 53, no. 11, pp. 1313–1317, Nov. 2006.

[18] S. Minato, "Zero-suppressed BDD's for set manipulation in combinatorial problems," in *Proc. 30th IEEE/ACM Design Automation Conf.*, Dallas, TX, 1993, pp. 272–277.

[19] G. Shi, "Computational complexity analysis of determinant decision diagram," *IEEE Trans. on Circuits and Systems - II: Express Briefs*, vol. 57, no. 10, pp. 828–832, 2010.

[20] ——, "A simple implementation of determinant decision diagram," in *Proc. International Conf. on Computer-Aided Design (ICCAD)*, San Jose, CA, USA, Nov. 2010, pp. 70–76.

[21] X. D. Tan and C. J. R. Shi, "Hierarchical symbolic analysis of analog integrated circuits via determinant decision diagrams," *IEEE Trans. on Computer-Aided Design of Integrated Circuits and Systems*, vol. 19, no. 4, pp. 401–412, April 2000.

[22] C. J. R. Shi and X. D. Tan, "Compact representation and eff cient generation of s-expanded symbolic network for computer-aided analog circuit design," *IEEE Trans. on Computer-Aided Design of Integrated Circuits and Systems*, vol. 20, no. 7, pp. 813–827, July 2001.

[23] W. Verhaegen and G. E. Gielen, "Eff cient DDD-based symbolic analysis of linear analog circuits," *IEEE Trans. on Circuits and Systems – II: Analog and Digital Signal Processing*, vol. 49, no. 7, pp. 474–487, 2002.

[24] S. X. D. Tan, W. Guo, and Z. Qi, "Hierarchical approach to exact symbolic analysis of large analog circuits," in *Proc. Design Automation Conference*, 2004, pp. 860–863.

[25] S. X. D. Tan and C. J. R. Shi, "Eff cient approximation of symbolic expressions for analog behavioral modeling and analysis," *IEEE Trans. on Computer-Aided Design of Integrated Circuits and Systems*, vol. 23, no. 6, pp. 907–918, June 2004.

[26] G. J. Minty, "A simple algorithm for listing all the trees of a graph," *IEEE Trans. on Circuit Theory*, vol. CT-12, p. 120, 1965.

[27] H. Xu, G. Shi, and X. Li, "Hierarchical exact symbolic analysis of large analog integrated circuits by symbolic stamps," in *Prof. Asia South-Pacific Design Automation Conference (ASPDAC)*, Yokohama, Japan, Jan. 2011, accepted for publication.

[28] A. Vladimirescu and S. Liu, "The simulation of MOS integrated circuits using SPICE2," EECS Department, University of California, Berkeley, Tech. Rep. UCB/ERL M80/7, 1980. [Online]. Available: http://www.eecs.berkeley.edu/Pubs/TechRpts/1980/9610.html

[29] W. J. McCalla, *Fundamentals of Computer-Aided Circuit Simulation*. Deventer, the Netherlands: Kluwer Academic Publishers, 1988.

[30] T. McConaghy and G. G. E. Gielen, "Globally reliable variation-aware sizing of analog integrated circuits via response surfaces and structural homotopy," *IEEE Trans. on Computer-Aided Design of Integrated Circuits and Systems*, vol. 28, no. 11, pp. 1627–1640, Nov. 2009.

[31] O. Guerra, E. Roca, F. V. Fernández, and A. Rodríguez-Vázquez, "Approximate symbolic analysis of hierarchically decomposed analog circuits," *Analog Integrated Circuits and Signal Processing*, vol. 31, pp. 131–145, 2002.

[32] A. Doboli and R. Vemuri, "A regularity-based hierarchical symbolic analysis methods for large-scale analog networks," *IEEE Trans. on Circuits and Systems – II: Analog and Digital Signal Processing*, vol. CAS-48, no. 11, pp. 1054–1068, 2001.

[33] J. Vlach and K. Singhal, *Computer Methods for Circuit Analysis and Design*. New York, NY: Van Nostrand Reinhold Company, 1983.

[34] J. Y. Lee, X. Huang, and R. A. Rohrer, "Pole and zero sensitivity calculation in asymptotic waveform evaluation," *IEEE Trans. on Computer-Aided Design*, vol. 11, no. 5, pp. 586–597, May 1992.

[35] X. Huang, C. S. Gathercole, and H. A. Mantooth, "Modeling nonlinear dynamics in analog circuits via root localization," *IEEE Trans. on Computer-Aided Design of Integrated Circuits and Systems*, vol. 22, no. 7, pp. 895–907, July 2003.

[36] L. T. Pillage and R. A. Rohrer, "Asymptotic waveform evaluation for timing analysis," *IEEE Trans. on Computer-Aided Design of Integrated Circuits and Systems*, vol. 9, no. 4, pp. 352–366, April 1990.

[37] G. Shi and X. Meng, "Variational analog integrated circuit design by symbolic sensitivity analysis," in *Proc. International Symposium on Circuits and Systems (ISCAS)*, Taiwan, China, May 2009, pp. 3002–3005.

[38] H. Yang, A. Agarwal, and R. Vemuri, "Fast analog circuit synthesis using multiparameter sensitivity analysis based on element-coeff cient diagrams," in *Proc. IEEE Computer Society Annual Symposium on VLSI*, Tampa, Florida, USA, 2005, pp. 71–76.

[39] H. Yang, M. Ranjan, W. Verhaegen, M. Ding, R. Vemuri, and G. Gielen, "Eff cient symbolic sensitivity analysis of analog circuits using element-coeff cient diagrams," in *Proc. Asia South-Pacific Design Automation Conference (ASPDAC)*, Yokohama, Japan, Jan. 2005, pp. 230–235.

[40] D. Ma, G. Shi, and A. Lee, "A design platform for analog device size sensitivity analysis and visualization," in *Proc. Asia Pacific Conference on Circuits and Systems (APCCAS)*, Malaysia, Dec. 2010, pp. 48–51.

[41] X. Li, H. Xu, G. Shi, and A. Tai, "Hierarchical symbolic sensitivity computation with applications to large amplifer circuit design," in *Prof. International Conference on Circuits and Systems (ISCAS)*, Rio de Janeiro, Brazil, May 2011, accepted for publication.

In: Analog Circuits: Applications, Design and Performance
Editor: Esteban Tlelo-Cuautle

ISBN: 978-1-61324-355-8
© 2012 Nova Science Publishers, Inc.

Chapter 2

DESIGN ISSUES OF SiGe HBT BASED ANALOG CIRCUITS

R. K. Chauhan[*]

INAOE, Department of Electronics and Communication Engineering
M.M.M. Engineering College, Gorakhpur-273010, INDIA

Abstract

The need to serve the relentless demand of high bandwidth required in the operation of analog circuits consisting of several discrete devices for various applications has driven transistor performance requirements beyond the reach of conventional silicon devices. As alternatives are failed to match the cost and other constraints of this uncompromising field, the industry is looking towards Silicon-germanium (SiGe) Heterojunction Bipolar Transistors (HBTs) as a cost-effective, process-compatible technology for the future devices and circuits. SiGe HBT technology combines transistor performance competitive with III–V technologies with the processing maturity, integration levels, yield, and hence, cost commonly associated with conventional Silicon (Si) fabrication. State-of-the-art SiGe HBT's can deliver high f_T, high f_{max}, low noise figure, better working at cryogenic operation, excellent radiation hardness, competitive power amplifiers and reliability comparable to Silicon. Chapter starts with the discussion on analog circuit design and issues related to it. Some of the important design parameter pertaining to SiGe HBT and its effect on the performance of the HBT based analog circuits is discussed thereafter.

Introduction

Analog circuits are of great importance in electronic system design since our physical world is fundamentally analog in nature. Therefore, the circuits designed for analog operation directly interact with the signals obtained or generated from the physical world. In the last two decade, the advancement of digital systems outpaces analog system design. Since most digital systems require analog modules for interfacing to the external world, it plays an important role in the creation of every modern IC. First and foremost, analog circuits act as

---

[*] E-mail address: rkchauhan27@gmail.com

the interface between digital systems and the real world. They act to amplify and filter analog signals, and to convert signals from analog to digital and back again. These analog interfaces appear in all communications devices (e.g., cell phones) - both to condition the "transmitted" signal and as sensitive "receivers." In addition, these analog interfaces appear in sensors (e.g., accelerometers). In addition, high performance digital cell design (either high speed or low power) also invokes significant analog circuit design issues.

In space applications, the automated design of analog circuitry could hold many benefits, especially for controller hardware. For example, in problems where actuator outputs need to be rapidly modulated in response to sensor feedback, analog circuits have some clear advantages over digital circuits. Specifically, digital control circuits necessitate a costly bandwidth-limiting analog to digital conversion (ADC) of the sensor signal and then a reverse conversion (DAC) from the processed digital signal to the analog actuator control output. These transformations lose information and introduce latency. In addition, the high frequency component of the sensor signal that is crucial to usefully controlling the actuators in many tasks is lost in the analog to digital conversion.

Computer aided design (CAD) methodologies have been very successful in automating the design of digital systems given a behavioral description of the function desired. Such is not the case for analog circuit design. Analog design still requires a "hands on" design approach in general. Moreover, many of the design techniques used for discrete analog circuits are not applicable to the design of analog/mixed-signal VLSI circuits. It is necessary to examine closely the design process of analog circuits and to identify those principles that will increase design productivity and the designer's chances for success. Thus, this chapter provides a hierarchical organization of the subject of analog integrated-circuit design and identification of its general principles.

Integrated circuits can be classified on the basis of signals used in the circuit as: Analog, digital or mixed signals circuits. To characterize the first two design methods, one must understand clearly the analog and digital signals. A signal can be in any of the following forms that can be detected: Voltage, current or charge.

Main requirement for any circuit design is that it must convey information about the state or behavior of a physical system. An analog signal is a signal that is defined over a continuous range of time and a continuous range of amplitudes. A digital signal is a signal that is defined only at discrete values of amplitude and it is quantized to discrete values. If circuit design comprises input and output signals that are analog in nature, it is analog circuits and similarly digital circuits are defined as the circuit comprising input/output signals which are digital in nature. Mixed signal circuits comprise both analog as well as digital circuits design issues. A brief comparison of analog circuits and digital circuits are listed in Table.1

Choosing whether to use an analog circuit or digital-circuit architecture depends largely on the design's ability to convert the signal. If the design can perform acceptable analog-to-digital and digital-to-analog conversion, digital-signal processing can usually provide a suitable transfer function. Accuracy and sampling rates of converters become the limiting factors of this strategy. Analog signal processing is more susceptible to noise and process variances. Digital implementations, though more complex in nature, uses a smaller die area.

Analog circuits can be designed from the viewpoint of hierarchy. Fig. 1 shows a vertical hierarchy consisting of devices, circuits, and systems. The device level effects on design are considered as the lowest level of analog circuit design. It is expressed in terms of device specifications, geometry, or model parameters for the design. The circuit level is the next higher level of design and can be expressed in terms of devices. The circuit design may include voltage and current relationships, parameterized layouts, and macromodels. The highest level of design is the systems level and can be expressed in terms of circuits. The systems level design generally includes mathematical or graphical descriptions, a chip floor plan, and a behavioral model.

Table 1. Comparison of Analog circuits and Digital circuits

S.No	Analog Circuits	Digital Circuits
1	Signals are continuous in amplitude and can be continuous or discrete in time	Signal are discontinuous in amplitude and time - binary signals have two amplitude states
2	Designed at the circuit level	Designed at the systems level
3	Components must have a continuum of values	Component have fixed values
4	Generally Customized	Standard
5	CAD tools are difficult to apply	CAD tools have been extremely successful
6	Requires precision modeling	Timing models only
7	Performance optimized	Programmable by software
8	Irregular block	Regular blocks
9	Difficult to route automatically	Easy to route automatically
10	Dynamic range limited by power supplies and noise (and linearity)	Dynamic range unlimited

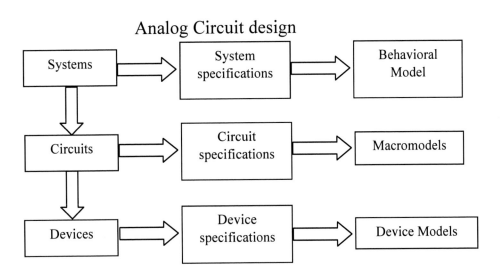

Figure 1. Flow chart showing analog circuit design.

Circuit design is the creative process of developing a circuit that solves a particular problem. Design can be better understood by comparing it to analysis. The analysis of a circuit, is the process by which one starts with the circuit and finds its properties. An important characteristic of the analysis process is that the solution or properties are unique. On the other hand, the synthesis or design of a circuit is the process by which one starts with a desired set of properties and finds a circuit that satisfies them. In a design problem the solution is not unique thus giving opportunity for the designer to be creative. Consider the design of a 1.5 µF capacitor as a simple example. This capacitance could be realized as the parallel connection of three 0.5 µF capacitor, the combination of a 1 µF capacitor in parallel with two 1 µF capacitors in series, and so forth. All would satisfy the requirement of 1.5 µF capacitance although some might exhibit other properties that would favor their use. Fig. 2 illustrates the difference between synthesis (design) and analysis.

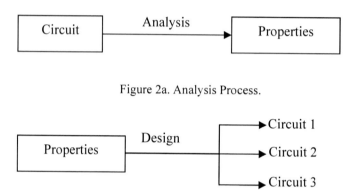

Figure 2a. Analysis Process.

Figure 2b. Design Process.

The circuit design process may involve the effect of different device parameters on a given circuit or different circuits with same device parameters. If the circuit design process involves the device parameter then characterization of device is necessary and its technology plays key role in the design. If the focus of this chapter is on only silicon technology then it can be classified as given in the Fig.3. The heterojunction bipolar transistor involves SiGe based HBT which is compatible to silicon technology easily.

The differences between integrated and discrete analog circuit design are important. Unlike integrated circuits, discrete circuits use active and passive components that are not on the same substrate. A major benefit of components sharing the same substrate in close proximity is that component matching can be used as a tool for design. Another difference between the two design methods is that the geometry of active devices and passive components in integrated circuit design are under the control of the designer. This control over geometry gives the designer a new degree of freedom in the design process. A second difference is due to the fact that it is impractical to breadboard the integrated-circuit design. Consequently, the designer must turn to computer simulation methods to confirm the design's performance. Another difference between integrated and discrete analog design is that the integrated-circuit designer is restricted to a more limited class of components that are compatible with the technology being used.

The task of designing an analog integrated circuit includes many steps. Some of the steps in the design process are listed as: Definition, Synthesis or implementation, Simulation or modeling, Geometrical description, Simulation including the geometrical parasitic, Fabrication, Testing and verification.

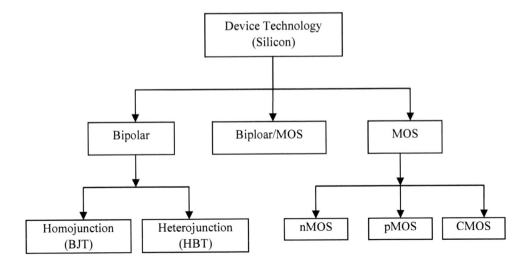

Figure 3. Silicon technology used in the fabrication of different devices.

The designer is responsible for all of these steps except fabrication. The first steps are to define and synthesize the function. These steps are crucial since they determine the performance capability of the design. When these steps are completed, the designer must be able to confirm the design before it is fabricated. The next step is to simulate the circuit to predict the performance of the circuit. The designer makes approximations about the physical definition of the circuit initially. Later, once the layout is complete, simulations are checked using parasitic information derived from the layout. At this point, the designer may iterate using the simulation results to improve the circuit's performance. Once satisfied with this performance, the designer can address the next step, the geometrical description (layout) of the circuit. As stated earlier, once the layout is finished, it is necessary to include the geometrical effects in additional simulations. If results are satisfactory, the circuit is ready for fabrication. After fabrication, the designer is faced with the last step, determining whether the fabricated circuit meets the design specifications. If the designer has not carefully considered this step in the overall design process, it may be difficult to test the circuit and determine whether or not specifications have been met.

As mentioned earlier, one distinction between discrete and integrated analog-circuit design is that it may be impractical to breadboard the integrated circuit. Computer simulation techniques have been developed that have several advantages, provided the models are adequate. These advantages include:

➤ Elimination of the need for breadboards
➤ Monitor signals at any point in the circuit
➤ Ability to open a feedback loop
➤ Easily modify the circuit.
➤ Analyze the circuit at different processes and temperatures

Disadvantages of computer simulation include:

➤ Dependence on the accuracy of models
➤ Failure of the simulation program to converge to a solution
➤ Time required for performing simulations of large circuits.
➤ Use of computer as a substitute for thinking.

In accomplishing the design steps, the designer works with three different types of description formats:

- ➢ the design description
- ➢ the physical description
- ➢ the model/simulation description

The format of the design description is the way in which the circuit is specified; the physical description format is the geometrical definition of the circuit; the model/simulation format is the means by which the circuit can be simulated. The designer must be able to describe the design in each of these formats. For example, the first steps of analog integrated-circuit design could be carried out in the design description format. The geometrical description obviously uses the geometrical format. The simulation steps would use the model/simulation format.

Disadvantages of computer simulation include:

- ➢ the accuracy of models.
- ➢ the failure of the simulation program to converge to a solution.
- ➢ the time required to perform simulations of large circuits.
- ➢ the use of the computer as a substitute for thinking.

Because simulation is closely associated with the design process, it will be included in the text where appropriate.

In accomplishing the design steps described above, the designer works with three different types of description formats. These include: the design description, the physical description, and the model/simulation description. The format of the design description is the way in which the circuit is specified; the physical description format is the geometrical definition of the circuit; the model/simulation format is the means by which the circuit can be simulated. The designer must be able to describe the design in each of these formats. For example, the first steps of analog integrated-circuit design could be carried out in the design description format. The geometrical description obviously uses the geometrical format. The simulation steps would use the model/simulation format.

The description of device generally used in analog circuits is given in Fig.3. Off these, BJT based and MOS based analog circuits are generally covered in every text book, therefore an attempt has been done here to describe the performance of analog circuits when HBT is used in place of BJT based circuits. The parameters that are essential in the design of HBT and CMOS based analog circuits are also covered in this chapter.

In the design of any analog circuits, the device technology must ensure certain requirements for its use such as its reliability, its speed together with current drive capability, high self-gain, low noise, and compatibility with Metal Oxide-Semiconductor (MOS) technology. Silicon-Germanium Heterojunction-Bipolar-Technology (SiGe-HBT) based circuit posses all those qualities which are listed above and therefore can be considered as a choice for many demanding applications.

Amplifiers are an integral part of most communication systems. The operation frequency of the communication systems like cell phones and LANs vary from 900MHz to 5.5GHz. For satellite communications the operating frequency is about 10 GHz. Hence the amplifiers are required to work well at these frequencies. While Silicon BJT works well for frequencies less than 2GHz, for high frequencies, III-V based transistors are used. Apart from being more expensive than the Si based technology, the III-V's cannot be integrated into Si based CMOS technology. Hence it is necessary to develop an amplifier that can work at higher frequency and that can be integrated into Si technology.

Bipolar transistors achieve their speed through the use of a very thin base region. The speed of the collector current depends on how long it takes for charge carriers to travel through the base. By introducing Ge into the base of a Si Bipolar Junction Transistor (BJT), a smaller base band gap is created, which increases electron injection from the emitter and thus decreases the base transit time. The decrease in transit time results in a higher f_T and higher β. A higher β permits the base doping level to be raised, lowering the base resistance. Research has found that the best performance is achieved by having the largest amount of Ge near the base-collector junction and the smallest amount near the base-emitter junction . This design maximizes the drift field for electrons in the base.

A broadband Darlington amplifier using SiGe-HBT can be designed and evaluated to achieve high gain at microwave frequency range. With careful design and development, the SiGe HBTs based Darlington Amplifier has the potential to replace III-V transistors. SiGe HBTs are cost effective and can be integrated into Si technology, thus making system on chip a reality. The gain, switching time and effect of temperature on a SiGe HBT based Darlington amplifier can be studied in an effort to improve its high frequency performance.

There are different ways to analyze the SiGe HBT based circuits. One of the approaches includes impact of Ge profile in the characterization of SiGe HBT and thereby its effects on the circuits based on it. The process requires characterizing and modeling the response of SiGe HBTs for different Ge profiles as a first step and then modeling and prediction of the effects of Ge profiles on the HBT based circuit can be worked out. Hence, the whole work of analyzing any analog circuit involving SiGe HBT through the above mentioned approach can be summarized as:

(i)   Develop an analytical model to study the performance characteristics of the SiGe HBTs for the various Ge profiles.

(ii) Implementation of such model parameters in SPICE to simulate the full circuit response and see the impact of design parameters on the performance of circuits.

2. SiGe HBT and Its Characterization

The schematic layer structure of a SiGe-HBT is shown in Fig.4a. The structure shown in this figure is suitable for a monolithic integrated circuit comprising HBT. The base of the HBT is formed by the graded band-gap SiGe layer, where the grading depends on the Ge content incorporated into Si layer. A simple device structure of an n-p-n SiGe HBT has been considered as shown in Fig. 4b. Quite different from the normal bipolar junction transistor (where the emitter region is heavily doped) HBT considered here have a heavily doped base region. This rearrangement in doping is done to improve base resistance and speed performance without degrading the current gain.

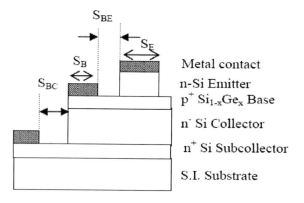

Figure 4a. Structure of an n-p-n SiGe HBT.

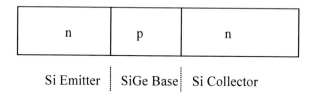

Figure 4b. Simplified structure of an n-p-n SiGe HBT.

2.1. DC Characteristics

The dc characteristics can be obtained only after understanding the consequences of introducing Ge into the base region of an SiGe HBT. To start with, one considers an energy-band diagram of the resultant device and compares it with that of Si based device.

2.1.1. Bandgap and Bandgap Narrowing

The cross-section of the SiGe HBT base is shown in Fig (4b). The bandgap of SiGe HBT is given by [13],

$$E_g^{SiGe} = E_g^{Si} \bullet (1-x) + E_g^{Gei} \bullet x + C_g \bullet (1-x) \bullet x \qquad (1)$$

Where x is Ge conc. and C_g = -0.4 eV is a bowing parameter. Ge is compositionally graded from the emitter-base (EB) junction with lower concentration to the collector–base (CB) junction with higher concentration and it is shown as dashed line in Fig (5).

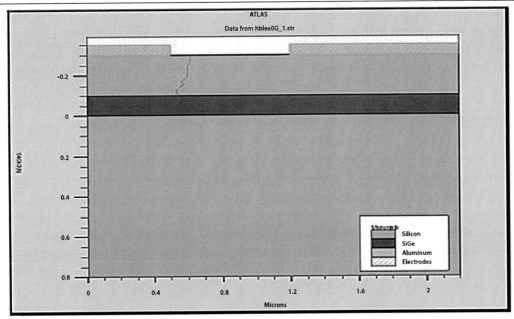

Figure 5. SiGe HBT used for simulation.

Due to this impact, the SiGe HBT consists of a finite band offset at the EB junction and a larger band offset at the CB junction. Bandgap grading is easily used for position dependence of the band offset with respect to Si. An electric field is produced by such position dependence in the Ge induced band offset in the neutral base region. This effect aids the transportation of minority carriers (electrons) from emitter to collector, which in turn improve the frequency response.

Thus we can say apparently that the Ge-induced reduction in base bandgap takes place at the EB edge of the quasi-neutral base $\Delta E_{g,Ge}$ (x=0) along with the CB edge of the quasi-neutral base $\Delta E_{g,Ge}$ (x=W$_b$). Thus the graded bandgap is given by equation as [14],

$$\Delta E_{g,Ge}(grade) = \Delta E_{g,Ge}(W_b) - \Delta E_{g,Ge}(0) \qquad (2)$$

It is assumed that the doping induced bandgap narrowing (bgn) in SiGe is identical to Si. the doped SiGe helps to calculate the bandgap of intrinsic SiGe by subtracting the doping induced bgn as [5],

$$E_g^{SiGe}(meV) = E_g^{Si}(meV) - 750x + 238x^2 \quad \text{for } Si_{1-x}Ge_x \qquad (3)$$

Here x is the mole fraction of Ge in Si-base of HBT. Assuming the T dependence of the SiGe bandgap similar to that of Si. Therefore, it was possible to verify by calculating the bandgap of SiGe at different T. The temperature reliant bandgaps of the components, E_g^{Si} and E_g^{Ge} are calculated by universally used Varshni model [7],

$$E_g = E_{g,0} - \frac{\alpha T_L^2}{\beta + T_L} \quad (4)$$

Where $E_{g,0}$ is the bandgap at $T_L = 0K$. The values of parameters are summarized in Table-2. The dependence on the material composition x is then pioneered by

$$E_g^{SiGe} = E_g^{Si} \bullet (1-x) + E_g^{Ge} \bullet (x) + C_g \bullet (1-x) \bullet x \quad (5)$$

As shown in Fig. 6 by the dashed line, the Ge is compositionally graded from low concentration at the EB junction to high concentration at the collector–base (CB) junction. The impact on the SiGe HBT band diagram is also shown and consists of a finite band offset at the EB junction $\left|\Delta E_{g,Ge}(x=0)\right|$ as well as a larger band offset at the CB junction $\left|\Delta E_{g,Ge}(x=W_b)\right|$. The position dependence of the band offset with respect to Si is conveniently expressed as a bandgap grading term $\left|E_{g,Ge}(grade) = \Delta E_{g,Ge}(W_b) - \Delta E_{g,Ge}(0)\right|$. Physically this position dependence in the Ge induced band offset produces an electric field in the neutral base region which aids the transport of minority carriers (electrons) from emitter to collector, thereby improving frequency response.

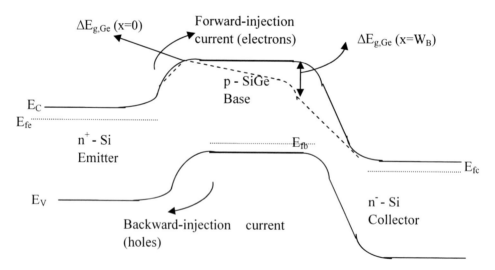

Figure 6. Energy Band Diagram for a Si BJT. Dotted line shows graded-base SiGe HBT.

Table 2. Parameter values for modeling the Bandgap Energy

Material	$E_{g,0}(eV)$	$\alpha\left(\dfrac{eV}{K}\right)$	$\beta(K)$
Si	1.1695	4.73×10^{-4}	636
Ge	0.7437	4.774×10^{-4}	235

2.1.2. Collector Current

The important dc consequence of adding Ge into the base, however, lies with the collector current density (J_C). Physically, the barrier to electron injection at the EB junction is reduced by introducing Ge into the base, yielding more charge transport from emitter-to-collector for a given applied EB bias. Observe from Fig. 5 that in this graded-base design, the emitter region of the SiGe HBT and Si BJT comparison are essentially identical, implying that the resultant base current density (J_B) of the two transistors will be roughly the same. The net result is that the introduction of Ge increases the current gain ($\beta = J_C / J_B$) of the transistor. From a more device–physics oriented viewpoint, the Ge-induced band offset exponentially decreases the intrinsic carrier density in the base which, in turn, decreases the base Gummel number and, hence, increases. Meaningful in this context is the Ge-induced improvement in current gain over a comparably constructed Si BJT and can be described as [1]

$$\frac{\beta_{SiGe}}{\beta_{Si}} = \gamma\eta \frac{\Delta E_{g,Ge}(grade)/kTe^{\Delta E_{g,G}(0)/kT}}{1-e^{\Delta E_{g,G}(grade)/kT}} = \frac{\tau_{B,SiGe}}{\tau_{B,Si}}$$

$$= \frac{2}{\eta} \bullet \frac{kT}{\Delta E_{g,Ge}(grade)} \bullet \left\{ 1 - \frac{kT}{\Delta E_{g,Ge}(grade)} \left[1 - e^{-\Delta E_{g,Ge}(grade)/kT} \right] \right\} \qquad (6)$$

And η is the electron diffusivity ratio between SiGe and Si as [14],

$$\eta = \frac{(D_{nb})_{SiGe}}{(D_{nb})_{Si}} \qquad (7)$$

And an "effective density-of-states ratio" between SiGe and Si according to [14]

$$\gamma = \frac{(N_C N_V)_{SiGe}}{(N_C N_V)_{Si}} < 1 \qquad (8)$$

Thus built-in field increases on decreasing the base width so the transit time decreases on higher built-in field ξ and higher collector current I_c. Here W_B is the width of the quasi-neutral base; D_{nb} is the diffusion coefficient of the minority- carrier electrons in the base.

As can be seen in Fig. 7, which compares the measured Gummel characteristics for two identically-constructed SiGe HBT's and Si BJT's, these theoretical expectations are clearly borne out in practice. The exponential dependence of β in an SiGe HBT on the EB boundary value of the Ge-induced band offset provides a powerful lever for tailoring β for a specific need and, importantly, effectively decouples β from the specifics of the base doping profile. For instance, in applications which do not require large values of β, the base can be more heavily doped to reduce the base resistance R_b without a negative impact on β, as would be the case in an Si BJT. That is, we can trade β for lower R_b to yield enhanced frequency response and better broad-band noise performance. This Ge lever can also be used to engineer an SiGe HBT whose β is independent of temperature [58], a decided advantage over Si BJT's from a circuit standpoint.

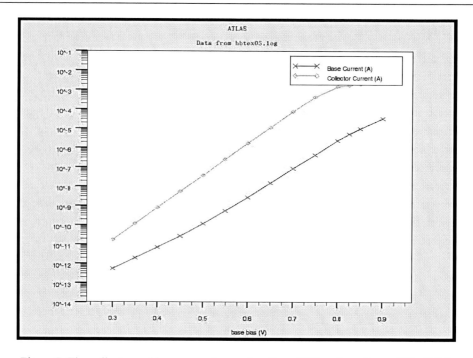

Figure 7. The collector and base currents as a function of EB voltage for an SiGe HBT.

2.2. Frequency Response

In most RF and microwave circuit applications, it is the transistor frequency response that limits system performance. An important figure of merit in bipolar transistors is the unity–gain cutoff frequency (f_T), which is given by[1]

$$f_T = \left[\frac{1}{g_m}(C_{eb} + C_{cb}) + \tau_b + \tau_e + \tau_c\right]^{-1} \qquad [9]$$

Where g_m is the transconductance, C_{eb} and C_{cb} are the EB and CB capacitances, and τ_b, τ_e and τ_c are the base, emitter, and collector transit times, respectively. The f_T of a transistor in principle samples only the vertical profile, and is thus a useful metric for comparing various technologies. In conventional Si BJT's, τ_b typically limits the maximum f_T of the transistor. The built-in electric field induced by the Ge grading across the neutral base effectively decreases τ_b since, physically, the carriers are more rapidly accelerated across the base. With respect to an identically constructed Si BJT we find [1]

$$\frac{\tau_{B,SiGe}}{\tau_{B,Si}} = \frac{2}{\eta} \cdot \frac{kT}{\Delta E_{g,Ge}(grade)} \cdot \left\{1 - \frac{kT}{\Delta E_{g,Ge}(grade)}\left[1 - e^{-\Delta E_{g,Ge}(grade)/kT}\right]\right\} \qquad [10]$$

Design Issues of SiGe HBT Based Analog Circuits

In addition, since τ_e is reciprocally related to the ac β of the transistor, the band offset at the EB junction also serves to improve the SiGe HBT frequency response [14].

$$\frac{\tau_{B,SiGe}}{\tau_{B,Si}} = \frac{\beta_{SiGe}}{\beta_{Si}} = \gamma\eta \frac{\frac{\Delta E_{g,Ge}(grade)}{kT} e^{\frac{\Delta E_{g,Ge}(0)}{kT}}}{1 - e^{-\Delta E_{g,G}(grade)/kT}} \qquad [11]$$

While τ_e is not typically a limiting transit time in state-of-art polysilicon-emitter devices, it will become increasingly important as the vertical profile is scaled to thinner dimensions with further technology evolution.

The unity power-gain frequency (f_{max}), or maximum oscillation frequency, is a more relevant figure of merit for practical RF and microwave applications since it depends not only on the intrinsic transistor performance (f_T), but also the parasitics of the device, according to [1]

$$f_{max} = \sqrt{\frac{f_T}{8\pi \, C_{CB} R_B}} \qquad [12]$$

Where C_{BC} is the base-collector junction capacitance and R_B is the base resistance. Thus from equation (8) it is comprehensible that lower base-collector junction capacitance and base resistance are required for the higher value of f_{max}. So the cut-off frequency f_T is increases as transit time decreases which in turn affect the f_{max}.

2.3. High Frequency Characterization

High frequency devices are characterized by several methods. These methods include the scattering parameter(S-parameter) measurement by means of a network analyzer. S-parameter measurement is a mean of a small signal measurement. The main issue in this measurement is the calibration with accurately de-embedding the parasitic. The load-pull measurement, a very useful technique is used for power measurements, even though it is costly and time consuming. A set of two-port parameters [including Z-, Y-, H-, ABCD-, and S- parameters] are used to describe the small signal RF performance of SiGe HBT. One type of parameters can simply be converted into another type of parameters with the help of matrix manipulation. The Y- parameters are frequently most suitable for equivalent circuit based analysis whereas the s-parameters are almost absolutely used for RF as well as microwave measurements due to practical reasons [20].

2.3.1. S-parameters

The device under test (DUT) often oscillates by open and short terminations because at very high frequencies, accurate open and short circuits are not easy to attain due to inherent parasitic capacitances and inductances. The interconnection between the DUT and test apparatus is also akin to the wavelength. The consideration of distributive effects is required in this case. So S-parameters were developed and are almost absolutely used to portray transistor RF with Microwave performance due to these practical difficulties.

S- parameters have almost similar information like Z-, Y-, or H-parameters. But the difference between them is that the dependent and independent variables are no longer simple voltages and the currents in S-parameters. In addition, in the s-parameter four "voltage waves" are produced by linear combinations of the simple variables. These voltage waves hold the identical information because they are chosen to be linearly independent. Transmission line techniques are used to compute s-parameters at high frequencies by properly selection of these combinations.

If incident wave is indicated by 'a' and reflection or scattering is indicated by 'b'. Here the voltage waves are defined using voltages and currents for characteristic impedance Z_0. Thus voltages (a_1, a_2) are called incident waves, while (b_1, b_2) are called scattered waves. A set of linear equations is used to relate the scattered waves with the incident waves by just as the port voltages are related to the port currents by Z-parameters.

It is expressed like,

$$\begin{pmatrix} b_1 \\ b_2 \end{pmatrix} = \begin{pmatrix} S_{11} & S_{12} \\ S_{21} & S_{22} \end{pmatrix} \begin{pmatrix} a_1 \\ a_2 \end{pmatrix} \tag{13}$$

Where, S_{11} is simply reflection coefficient corresponding to input impedance with Z_0 output termination. S_{22} is output reflection coefficient looking back into the output for Z_0 source termination.

$|S_{21}|^2$ is transducer gain for a Z_0 source and Z_0 load.

$|S_{12}|^2$ is reverse transducer gain for a Z_0 source and a Z_0 load. Thus S-parameters of a SiGe HBT will thoroughly depend on the size of transistor, frequency of operation as well as on biasing condition.

2.3.2. De-Embedding

De-Embedding is the process for removing; de-embed (the surrounding parasitic). This is done for characterizing only intrinsic part of the device. There are numerous ways to de-embed the parasitic with diverse levels of precision and complication. The first simple and directly way is to apply lumped circuit model for representing the parasitic. Secondly, compose device under test with different width by using identical metal pattern. When it is done then admittance is used to estimate the parasitic. This admittance is achieved when the width is extrapolated to zero. Furthermore, relate the similar calibrations routines and these routines are built-in VNA. Thus De-Embedding process can be completed by two ways; initially by using the calibration structures during calibrating the network analyzer and secondly by calibration on standard ISS (impedance standard substrate).

2.4. Small-signal HF Modeling and Performance Factors

In this section we develop a novel and uncomplicated extraction method for discussing the transistor RF performance along with procedures to find out the parameters of SiGe HBT by means of small-signal Π topology equivalent circuits of this HBT. The algorithm is helpful for extracting both intrinsic plus extrinsic (parasitic) elements. If we determine formerly the extrinsic elements of the HBT then conventional procedures or methods derived from simple bias measurements work very sound. Through different procedures for example DC, cut-off measurements, or optimization can be used for this approach. Since the typical DC and cut-off techniques present poor performance for Silicon Germanium HBT devices that's why it is frequently very hard to precisely determine the values of parasitic elements of the HBT. An innovative technique has been developed to circumvent this problem and in this technique only scattering (S)-parameters at different biases are measured. For fitting the measured S-parameters appropriately, linear models by way of a Π topology have been experienced. We have neglected emitter resistance, the collector resistance, along with the output resistance due to Early effect for simplicity [17].

S-parameters obtained from ac analysis are simply converted into Y-, Z- or H-parameters using ATLAS. Various Power Gains for example MAG (maximum available gain), MSG (maximum stable gain) as well as MAUG (maximum available unilateral gain) are used for analysis. Furthermore, a figure-of-merit that has been used extensively for microwave characterization is MSG. Due to simplicity of measurement at high frequencies these quantities are calculated from the measured small-signal scattering parameters.

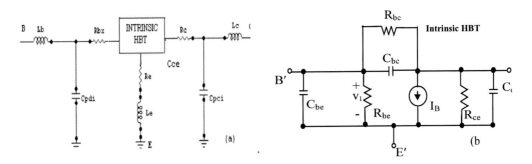

Figure 8. Schematic of a small-signal Π equivalent circuit of an HBT device. (a) It contains intrinsic and extrinsic circuit elements. (b) It contains the intrinsic elements and can be determined from the admittance parameters of the device at a number of different bias points.

The maximum stable gain is calculated by y_{21} and y_{12} as,

$$MSG = \left| \frac{y_{21}}{y_{12}} \right| \qquad (14)$$

And the maximum available gain is extracted as,

$$MAG = \left|\frac{y_{21}}{y_{12}}\right| \left(k - \sqrt{k^2 - 1}\right) \tag{15}$$

Where k is **'Rollett stability factor'** and extracted by this equation as,

$$k = \frac{2\,\mathrm{Re}(y_{11})\mathrm{Re}(y_{22}) - \mathrm{Re}(y_{12}y_{21})}{|y_{12}y_{21}|} \tag{16}$$

Mansion's gain is obtained by the following equation as,

$$U = \frac{|y_{21} - y_{12}|^2}{4[\mathrm{Re}(y_{11})\mathrm{Re}(y_{22}) - \mathrm{Re}(y_{12})\mathrm{Re}(y_{21})]} \tag{17}$$

The maximum available unilateral gain is calculated by this equation as,

$$MAUG = \frac{|y_{21}|^2}{4\,\mathrm{Re}(y_{11})\mathrm{Re}(y_{22})} \tag{18}$$

When both input and output are concurrently conjugate matched, at this state we achieve MAG. When k > 1, at this the device is unconditionally stable and MAG exists. It is obvious from equations (13) and (14), if the device is unilateral ($y_{12} = 0$) then U equals to MAUG. When the device is unconditionally stable then MAG equals to MSG and vice-versa. Maximum frequency at which MSG becomes unity is frequently termed as f_{max}. As power gain with no impedance transformation is achieved by common-emitter microwave transistors. This is the reason why these transistors may comprise useful gain when inserted into a system with 50 Ω. For this model, MSG (maximum stable gain) is called FOM (figure of merit). This device is unconditionally stable here.

f_T does not consider the limitations due to physical base resistance R_B but it is usually used to depict the frequency response of a bipolar transistor. Moreover another parameter named maximum oscillation frequency f_{max} includes the effect of R_B, with the parameters in f_T. The maximum oscillation frequency f_{max} is the extrapolated frequency at which the small signal power gain is reduced to unity when the terminations are conjugatly matched at both input and output. From the extrapolation of the high-frequency asymptote of a plot of the magnitude of h_{21} in dB versus log (frequency), we can extract the unity gain cut-off frequency. At amply low frequency most bipolar transistor devices may be depicted as single pole devices. This theory is alike to a high-frequency asymptote with a slope of -20 dB per decade so is the cut-off frequency though both C_{BE} and C_{BC} capacitances are bias dependent. f_{max} is extracted at the point where MSG becomes 0 dB from MSG (in dB) versus log (frequency) plot [17-18].

We can achieve the intrinsic elements from the admittance parameters of the HBT device for each bias point after a usual de-embedding process for taking the consequence of parasitic elements. The intrinsic elements for example C_{BE} (base-emitter capacitance) vary with frequency. The intrinsic part of the device (Fig. 2b) is contained by the given circuit topology of the small-signal model which is shown by dashed box in Fig. (2a). For a number of different devices tested, the estimated values of the pad parasitic capacitances C_{pbi} along with C_{pci} do not surpass hundredths of femto Farads (fF) due to this fact these capacitances can be omitted. The intrinsic and extrinsic parameters in Fig. 2 can be extracted by the following method [16]:

The base-emitter junction capacitance is calculated by the equation as follow [16],

$$C_{BE} = \frac{\mathrm{Im}[Y_{11}] + \mathrm{Im}[Y_{12}]}{\omega_i} \qquad (19)$$

The base-collector junction capacitance is expressed by the equation as follow [16],

$$C_{BC} = \frac{-\mathrm{Im}[Y_{12}]}{\omega_i} \qquad (20)$$

The base-collector junction capacitance is expressed by the equation as follow [16],

$$C_{CE} = \frac{\mathrm{Im}[Y_{22}] + \mathrm{Im}[Y_{12}]}{\omega_i} \qquad (21)$$

The base-collector junction capacitance is calculated by the equation as follow [16],

$$R_{BC} = \frac{-1}{\mathrm{Re}[Y_{12}]} \qquad (22)$$

The collector-emitter junction resistance is obtained by the equation as follow [16],

$$R_{CE} = \frac{1}{\mathrm{Re}[Y_{12}] + \mathrm{Re}[Y_{22}]} \qquad (23)$$

The collector-emitter junction resistance is expressed by the equation as follow [16],

$$R_{BE} = \frac{1}{\mathrm{Re}[Y_{11}] + \mathrm{Re}[Y_{12}]} \qquad (24)$$

And extrinsic resistance is obtained by the equation as follow [16],

$$R_{EXTRINSIC} = Z_{11} - Z_{12} \qquad (25)$$

In this method C_{BE} are intrinsic junction capacitances and R_{BC}, R_{CE} and R_{BE} are intrinsic junction resistances.

An analytical model of n-p-n SiGe HBT that takes into account the effect of Ge profiles on the performance of HBT in terms of its gain and its speed has been investigated. Other quantities, such as collector-junction capacitance, emitter-base capacitance are obtained from the model considered. The response of the individual transistor for various Ge profiles is then used to elucidate and predict the behavior of few analog circuits such as common emitter amplifier circuit, Darlington pair, active inductor involving current source.

3. Performance Analysis of Some Analog Circuits Consisting of SiGe HBT

3.1. Common Emitter Amplifier

The impact of Ge profiles in the base region of SiGe HBT has been studied on the real world common emitter amplifier circuit[23]. It was observed that the frequency response, current gain roll-off and gain bandwidth was altered with the shape and quantity of Ge addition into the base region. The gain bandwidth product (f_T), a major design issue in analog circuits, increases for all the profiles on increasing Ge concentration. Although triangular profile offers a high range of gain bandwidth product, but its DC gain is quite low. GDB and GSB profiles not only offer a significantly high range of f_T but also have a phenomenal DC current gain. These properties craft these profiles as a versatile option for designing the next generation devices. GDB and GSB profiles have been shown as the cost-effective options using Ge in the base region of SiGe HBT. These profiles will play a prominent role in future analog device domain where one requires a high f_T with high DC gain.

The HBT structure considered used for simulation has base width of 500Å and the doping in emitter, base and collector regions are 5×10^{16}, 10^{17} and 10^{14} per cm^3 respectively. SPICE parameters are extracted for all the profiles of HBT considered here for simulation of HBT based circuit. For various Ge profiles one can determine the performance of respective HBTs. Popularly two profiles are used, Box profile and Linearly graded profile. GDB and GSB profile were presented by Ankit and R K Chauhan [22]. In GDB profile, Ge addition in the first 20% and last 20% of the base region is done in Box shape, and for the remaining 60% of the base region, it is added in a linearly graded manner. For GSB profile whole base is divided into seven equal regions[22]. In alternate 4 regions, Ge is added in the box shape, while remaining regions has linearly graded Ge profile.

With the extracted SPICE parameters of HBTs, the Common emitter amplifier (shown in figure 9) is simulated in ORCAD. In all the results dash-dot line depicts Box profile while Dotted line is for triangular profile, and the two new profiles i.e. GDB and GSB are shown by darker and lighter lines respectively. Fig.10 shows the I-V characteristics of n-p-n SiGe HBT for different Ge profiles at the base current $I_B = 250\mu A$. All the profiles have the average Ge concentration of 12% in the base region. It has been found that GDB and GSB offer a much higher collector current for the same collector to emitter voltage drop in the circuit. In the active region, the collector current for the GDB and GSB profiles is found to be nearly 5 times that of triangular profile. This clearly indicates the importance of GDB and GSB over conventional profiles.

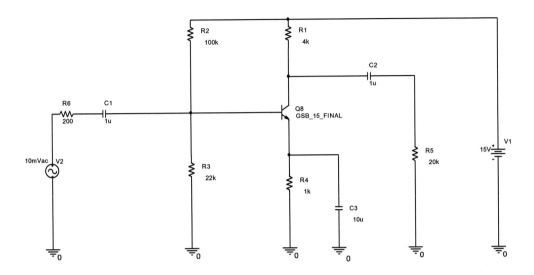

Figure 9. Schematic of CE amplifier used for simulation.

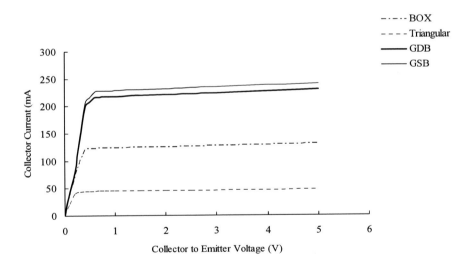

Figure 10. I-V Characteristics of HBT for various Ge profiles.

Fig.11 shows the frequency response of common emitter amplifier for different Ge profiles at 12% average Ge concentration. In this figure, we observed that at lower frequency range, except the triangular profiles, all the profiles offer a better output voltage level. But as the operational frequency increases, output voltage of triangular becomes highest compared to other profiles. It can be seen that GDB and GSB voltage gain performances are intermediate to the conventional profiles.

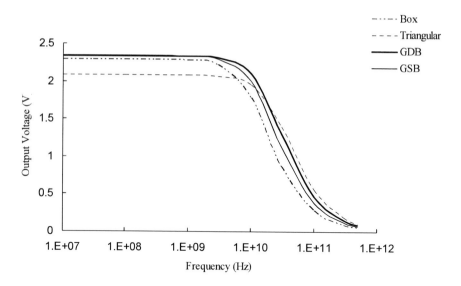

Figure 11. Frequency response of CE amplifier configuration for various Ge profiles.

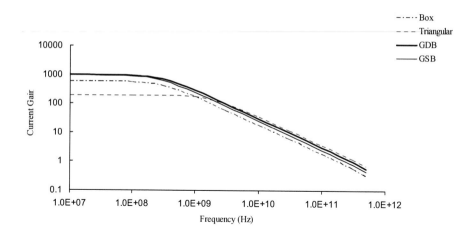

Figure 12. Current gain variation with change in frequency for various Ge Profiles.

Fig.12 depicts the current gain roll-off at higher frequencies for all the profiles at 12% average Ge concentration. As indicated in fig.10, the current gain is found to be quite high for these profiles at lower frequencies. On increasing frequency, roll-off is observed in all the profiles. It is observed that the performance of GDB profile is very close to the triangular. One good measure to evaluate the performance of profiles at this point is Gain bandwidth product (f_T), which is the frequency at which transistor offers unity current gain. The values of f_T obtained at 12% average Ge concentration for all the profiles are shown in Table 3.

Table 3. Gain bandwidth product for different profiles at 12% average Ge Concentration

Ge Profiles	Box	Triangular	GDB	GSB
f_T (GHz)	180	357	296	242

Darlington Pair:

The small base resistance of SiGe HBT's would minimize the chances of oscillation in a circuit involving fast clock operation. The reason is that the output inductance of the emitter follower is directly proportional to the base resistance of the emitter follower circuit. This, in turn, makes the resonance between the follower load capacitance and output inductance shift to a higher frequency.

The complete circuit schematic diagram of HBT based Darlington pair used in simulation is shown in Fig. 13. There is dc blocking capacitors at both input and output ports.

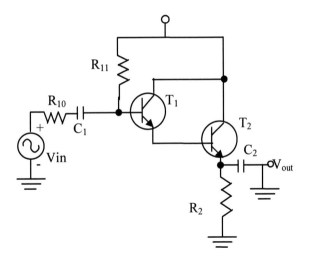

Figure 13. Schematic of high gain Darlington amplifier.

All the spice parameters for the SiGe HBT are extracted for the 3 to 15% linear doping of the Ge at base region [22]. Other values used in circuit simulation are shown in Fig.14. The simulated response of this circuit is shown in Fig. 2. It shows that the maximum gain max G of Darlington configurations using AlGaAs and SiGe is quite different. It is observed that the Darlington using SiGe improves the max G from dc to high frequency significantly compared to that of GaAs. The gain at 3 GHz is negative for GaAs where as it is 23dB for SiGe HBT based Darlington pair. The operation of SiGe based Darlington pair can operate satisfactorily till 23GHz. It can be said that circuit to be used in microwave range must be SiGe HBT

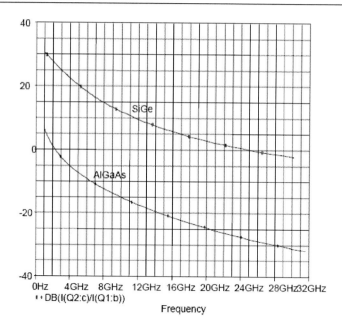

Figure 14. Response curve of GaAs and SiGe HBT based Darlington pair.

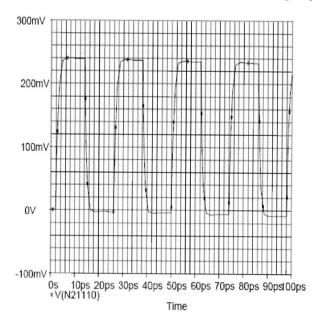

Figure 15. Switching response of SiGe HBT based Darlington circuit.

It was also seen that the switching time of such configuration is quite less. The switching time is shown in Fig. 15. It was found that the switching time is less than the conventional Darlington pair. Graph obtained from the simulation shows that the switching time is 5ps for SiGe Darlington pair configuration.

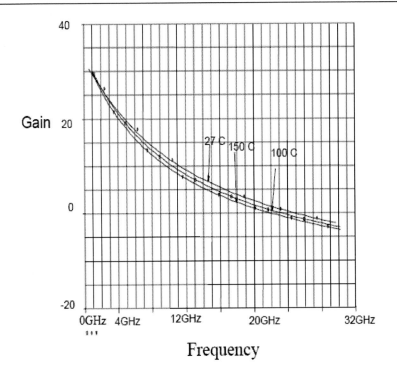

Figure 16. Variation in gain of SiGe Darlington with Temperature.

Fig.16 shows the response variation with increase in temperature. Three gain curves for 27C, 100C and 150C are depicted in the figure. One can observe that the gain varies very slowly with the temperature at high frequency. In addition, this is very important for microwave application where heat dissipation forces the device to operate at very high temperature.

3.3. Active Inductor Circuit Using SiGe HBT

The realization of active inductor circuits using SiGe HBT has been shown by Torres and Freirre[21]. Fig.17 shown below were simulated with CADENCE software using the foundry library models [10]. These models have shown the possibility of using this technology up to millimetre waves [21].

$$R = \frac{1}{g_{m2}}; \; C = C_{\pi 2}; \text{ and } L = \frac{C_{\pi 1}}{g_{m1} g_{m2}}$$

The inductance "L" for the above circuit when simulated by Torres and Freirre [21], it shows to have dependence on bias voltage and it has a maximum value of 400 pH when V_{bias} = 0.9 V. Active inductor realized in this fig, is for V_{CC} = 2 V, I_{CC} = 1.413 mA, V_{bias1} = 2.28 V biasing HBT_1, I_{bias1} = 0.447 mA, V_{bias2} = 1.2 V biasing HBT_4 and HBT_3 (current mirror), I_{bias2} = 1.45 mA was, L(28.5GHz) = 318.8 pH with a quality factor Q(28.5GHz) = 106. The active inductor presents a maximum Q at f_{maxQ} = 28.1 GHz and has a resonance frequency (real impedance, ImZout(fress) = 0) at f_{ress} = 39GHz.

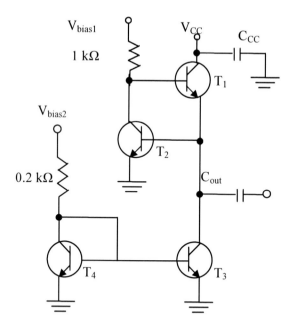

Figure 17. Active inductor circuit using SiGe HBT[12].

3.4. Variable Gain Amplifier

Monolithic SiGe heterojunction bipolar transistor (HBT) variable gain amplifiers (VGAs) with a feedforward configuration have been developed for 5 GHz applications by Chang-Wo Kim[25]. Two types of the feedforward VGAs were presented: one using a coupled-emitter resistor and the other using an HBT-based current source. The proposed VGA by Chang consists of four major blocks, an active balun, main differential amplifier, feedforward block, and control-voltage supplier as shown in Fig.18. The balun converts a single-ended input signal into two differential signals. These differential signals are amplified by the main amplifier and go out the RF out ports. The output signals of the VGA can be controlled by the gain of the main amplifier as well as the attenuation caused by the 180° out-of-phase feedforward signal. The amplified output signal is split to form two paths: one goes into the feedforward block and the other goes to a load resistor. The outgoing (180o out-of-phase) signal from the feedforward block is subtracted from the main signal. This subtraction process leads to the attenuation of the VGA. Consequently, the dB-linearity of the VGA improved easily provided the amount of the attenuation can be dB-linearly controlled by the control voltage. The results obtained by Chang were measured with a vector network analyzer and spectrum analyzer connected to a single ended output 50 Ω port. The collector bias voltage of V_{CC} used was of 3 V, and the control voltage were from 0 to 3 V. Results shows the measured gain control range at 5.2 GHz. The VGAs achieved a dynamic-gain-control ranges of 23.8 and 23.6 dB (–15 to 8.8 dB for the VGA with an emitter resistor and –15.4 to 8.2 dB for the VGA with a current source), respectively. The gain-control sensitivity of the VGAs were 90 mV/dB. In bipolar transistors, the gain-control voltages for 1 dB control should be much larger than the thermal voltage (26 mV/dB at room temperature) in order to be insensitive to temperature variation.

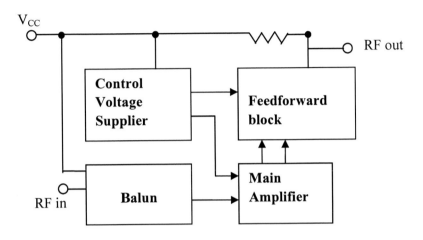

Figure 18. Feedforard VGA using SiGe HBT[25].

3.5. SiGe HBT in Optoelectronic Receiver Circuit

The design of SiGe based phototransistor can be done using the concept of photo-detector. The designs are based on a SiGe npn transistor layout. It is known that a complex layers are responsible for the formation of the SiGe base and afterwards the emitter directly above it. However, for a successful photo-detector a path is needed for light to reach SiGe base. Therefore the emitter had to be partially or fully removed. If the emitter is removed, the SiGe layer can be exposed to external light, creating a photodiode. On the other hand, if the base contact is removed and the emitter contact is maintained we create a phototransistor. Based on this idea several versions of photo-detectors were created.

Front-end transimpedance amplifiers (TIA) for optical receiver applications using the IBM 5HP (0.5micron) SiGe technology has been reported by Amit Gupta et al.[24]. This technology exhibits f_T and f_{MAX} of 47GHz and 65GHz respectively. Simulations in the Cadence Affirma analog design environment reported by them shows the TIA's output of 29mV for an input current of 10μA.

The schematic of TIA is shown in the fig.19 given below. It consists of the front end having common emitter amplifier, followed by a feedback resistance (R_f). This feedback is used to vary the input and output resistance of the first stage. npn transistors Q_1 and Q_2 is used as a common emitter stage followed by two cascaded common emitter stages Q_3 and Q_4 for high gain and bandwidth. The output stage consists of npn transistor Q_5 as a emitter follower which acts as a buffer.

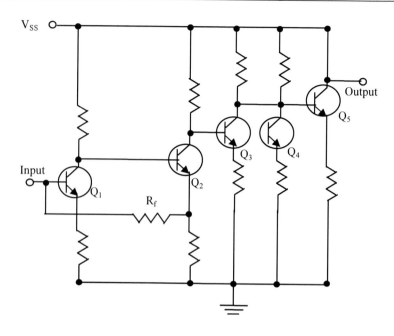

Figure 19. Application of SiGe HBT in TIA.

Summary

Now-a-days SiGe HBTs are comparable to the fastest III-V production devices in the high–speed orbit. The state-of-art in simulation of silicon germanium semiconductor devices is presented in this chapter. A comprehensive course of action to model the device parameter characterization of High Frequency 0.1μm SiGe HBT is depicted which is based on the technique of direct parameter extraction. With the help of S-, Y- and Z- parameters, the equivalent circuit parameters have been extracted. The intrinsic and the extrinsic elements of model are obtained using a direct extraction method that assists to find out the base resistance from the Z- parameters. The issues related to optimization of f_{max}, f_T and base resistance as well as junction capacitance are addressed.

Unique approach has been followed to analyze the SiGe HBT based circuits. The impact of Ge profile in the characterization of SiGe HBT and thereby its effects on the circuits based on it is covered. The process utilizes characterizing and modeling the response of SiGe HBTs for different Ge profiles as a first step and then modeling and prediction of the effects of Ge profiles on the HBT based circuit thereafter. Some of the important analog circuits used in communication equipment is shown utilizing SiGe HBT effectively.

Low cost and small size have been required for wireless communication terminals. Si based MMIC's have been focused as strategic devices to realize low production cost and high level of integration. SiGe technology is a leading contender when one requires monolithic active devices for ICs in the microwave range. Recently, SiGe HBT based VGA has been reported, and it shows the feasibility of a system on a chip. The experimental results reported by Chang shows that the developed VGAs can be a good candidate for wireless applications.

One of the proposed active inductors, aimed at millimetre wave band lower edge has been discussed on a low cost SiGe technology. The inductance value can be as high as 700pH for the cascode and almost 1nH for the CE+CC. Adjusting the bias conditions an inductance variation of 35% to 40% were reported by researchers. Discrepancies between measurements and simulation were due to parasitic capacitors underestimation. Presently the technology is used for implementing filters.

The optoelectronic application of SiGe has also been explored. Front-end transimpedance amplifiers (TIA) for optical receiver applications using the IBM 5HP (0.5micron) SiGe technology, reported by Amit Gupta et al.has also been discussed in the concluding part of this chapter.

References

[1] J.D. Cressler, SiGe HBT Technology: A New Contender for Si-Based RF and Microwave Circuit Applications, *IEEE Transc. Microwave Theory and Techniques*, vol. 46 (1998), 572.

[2] H. Kroemer, Heterojunction bipolar transistors and integrated circuits, *Proc. IEEE*, **70** (1982), 13.

[3] B.G. Streetman and S.K. Banerjee, Solid State Electronic devices 2005, PHI Publications.

[4] V.S. Patri and M. Jagdish Kumar, Novel Ge-profile designs for high-speed SiGe HBTs: modeling and analysis, *IEE Proc. Circuits Devices and System*, vol.146, No.5 (1999), pp. 291-296.

[5] M.K. Das, N.R. Das and P.K. Basu, Performance Analysis of a SiGe/Si Heterojunction Bipolar Transistor for Different Ge-composition , *Proceedings of the XXVIIIth UNION RADIO-SCIENTIFIQUE INTERNATIONALE (URSI)* General Assembly in New Delhi (Oct, 2005), Commission : D03.

[6] K.H. Kwok and C.R. Selvakumar, Profile Design Considerations for Minimizing Base Transit Time in SiGe HBTs for All Levels of Injection Before Onset of Kirk Effect, *IEEE Trans. Electron Devices*, vol. 48 no. 8 (2001), 1540.

[7] Z.M. Krstelj, V. Venkataraman, E.J. Printz, J.C. Sturm and C.W. Magee, Base resistance and effective bandgap reduction in n-p-n $Si/Si_{1-x}Ge_x/Si$ HBT's with heavy base doping, *IEEE Trans. Electron Devices*, vol. 43 (1996), 457.

[8] D.M. Richey, J.D. Cressler and A.J. Joseph, Scaling issues and Ge profile optimization in advanced UHV/CVD SiGe HBTs, *IEEE Trans. Electron Devices*, vol. 44 (1997), 431.

[9] H. Kroemer, "Two integral relations pertaining to the electron transport through a non uniform energy gap in the base region," *Solid State Electron.*, **28** (1985), 1101.

[10] K. Suzuki and N. Nakayama, "Base transit time of shallow base bipolar transistors considering velocity saturation at base-collector junction," *IEEE Trans. Electron Devices*, **39** (1992), 623.

[11] N. Jiang and Z. Ma, Current gain of SiGe HBTs under high base doping Concentration, *Semicond. Sci. Technol.* **22** (2007), S168.

[12] Jorge Alves Torres and J. Costa Freire, "Millimeter wave SiGe HBT voltage controlled active inductors," pp.1-8, APMC-2008.

[13] V. Palankovski and S. Selberherr, "Critical modeling issues of SiGe semiconductor devices", *J. Telecommun. Inform. Technol.*, no. 1, pp. 15{25}, 2004.

[14] John D. Cressler," Emerging SiGe HBT Reliability Issues for Mixed-Signal Circuit Applications" *IEEE Transactions on Device and Materials Reliability*, vol. 4, no. 2, June 2004.

[15] Zhenqiang Ma and Ningyue Jiang, "Base–Region Optimization of SiGe HBTs for High-Frequency Microwave Power Amplification", *IEEE Transactions on Electron Devices,* vol. 53, NO. 4, APRIL 2006

[16] J.M. Zamanillo, A. Tazon,A. Mediavilla and C. Navarro," Simple Algorithm Extracts SiGe HBT Parameters", MICROWAVES & RF, OCTOBER 1999.

[17] *ATLAS User's Manual DEVICE SIMULATION SOFTWARE*, SILVACO International,2004.

[18] Zhenqiang Ma et.al. "A High-Power and High-Gain X-Band Si/SiGe/Si Heterojunction Bipolar Transistor", *IEEE Transactions on Microwave Theory and Techniques*, vol. 50, NO. 4, APRIL 2002.

[19] Peter Ashburn," SiGe Hetrojunction Bipolar Transistors", Jhon Wiley & Sons Publication, 2003.

[20] Lars Vestling," Design and Modeling of High-Frequency LDMOS Transistors" *Comprehensive Summaries of Uppsala Dissertations from the Faculty of Science and Technology* **681**, ACTA. Universitatis UPSALIENSIS UPPSALA 2002.

[21] Jorge Alves Torres and J. Costa Freire, "Millimeter wave SiGe HBT voltage controlled active inductors," pp.1-8, APMC-2008.

[22] Ankit Kashyap and R.K. Chauhan, "A New Profile Design for Silicon Germanium based Hetero-Junction Bipolar Transistors," *Journal of Computational and Theoretical Nanosciences* (U.S.A.),Vol.5 (11), pp. 2238-2242, Nov 2008.

[23] Ankit Kashyap and R.K. Chauhan, "Effect of Ge Profile Design on the Performance of an n-p-n SiGe HBT Based Analog Circuits," *Microelectronics Journal (U.K.)*, **39**(12), pp. 1770-1773 , Dec 2008.

[24] Amit Gupta, Steven P. Levitan, Leo Selavo, Donald M. Chiarulli, "High-Speed Optoelectronics Receivers in SiGe," *Proceedings of 17th International Conf. on VLSI Design*, 2004.

[25] Chang-Woo Kim, "Monolithic SiGe HBT Feedforward Variable Gain Amplifiers for 5 GHz Applications" *ETRI*, Vol.28(3), pp. 386-388, 2006.

In: Analog Circuits: Applications, Design and Performance　　ISBN: 978-1-61324-355-8
Editor: Esteban Tlelo-Cuautle　　　　　　　　　　　© 2012 Nova Science Publishers, Inc.

Chapter 3

APPROXIMATION IN ANALOG SIGNAL PROCESSING

J. M. David Báez-López[*]
Departamento de Computación, Electrónica y Mecatrónica,
Universidad de las Américas-Puebla
Cholula, Puebla, 72820 MEXICO

Abstract

Analog signal processing circuits play an important role in modern circuit design because they can be u sed fo r signal con ditioning, filtering, preprocessing a,d postproc essing. They have been in use for about a centur y since the beginnings of wire less tran smissions and they are still useful in the realm of dig ital or di screte-time si gnal pro cessing. There are a n umber o f functions that signal processin g circuits can a ccomplish, but only a lim ited number of the m are widely known and used be cause they have proved to be t he most reliable and versatile in terms of circuit realization, sensitivity, and mathematical complexity.

In the de sign procedure fo r a sign al pr ocessing circu it, there are two steps: Approximation and synthesis. In the approximation step, which is the topic of this chapter, we produce a transfer function to be implemented by the circuit resulting from the synthesis step. The transfer function and its circuit realization must satisfy a set of specifications provided by the circuit de signer. We are only concerned, in this chapt er, with the requi rements that th e approximating function must satisfy.

In this chapter we make a detailed description of the most used approximation methods in analog signal processing circuits. They are: Butterworth, Chebyshev, Inverse, Chebyshev, and Thompson app roximations. The concept s of poles and zer os are exa mined in detail. In addition, the chapter makes a t reatment of delay in filter design. Finally, we cover frequency transformations to obtain high pass, band pass and band reject filters.

Introduction

This c hapter is de voted to the description a nd st udy of appr oximation t echniques for analog filter design. Appr oximation is the first step in the filter design proc ess. The second step is the implementation step or synthesis, where an actual circuit is designed to realize the

---

[*] E-mail address: dbaeziec@yahoo.com

approximation function. There are two types of approximation strategies. The most used one deals with the magnitude of the transfer function. In this case, we try to approximate the transfer function magnitude to an ideal brick-wall ideal magnitude. Depending upon the complexity of the proposed transfer function, the approximating function would be closer to the ideal magnitude characteristic. There are four types of magnitude approximation functions that we are going to cover in this chapter. They are the Butterworth approximation, which is a maximally flat magnitude characteristic and it can be obtained by making as many derivatives equal to zero at the origin and it has a monotonic response in the passband. The second magnitude characteristic that we describe is the Chebyshev one. This magnitude characteristic has an equal-ripple characteristic in the passband and it has a monotonic response in the stopband. The third one is the Inverse-Chebyshev characteristic, it has a passband with a monotonic response and an equal-ripple at the stopband. The last magnitude characteristic that we cover in this chapter is the elliptic characteristic that has equal-ripple at both the passband and the stopband. The second approximating strategy is a phase approximation. In this case we try to approximate the phase of an ideal delay filter. An ideal delay filter is a filter which only provides a delay. Of course, in addition we require a magnitude characteristic to be satisfied, but the emphasis is on the phase approximation. The phase approximation that we study here is the Thomson one.

Most of the chapter we will be mainly concerned with low pass functions. In the last section in the chapter we cover how to obtain high pass, band pass, and band reject functions.

Transfer Functions

A transfer function is a rational function in the complex variable s. It is the quotient of two polynomial that we call the numerator and denominator polynomials, Thus, a typical transfer function has the form

$$N(s) = H \frac{a_n s^n + a_{n-1} s^{n-1} + \cdots + a_1 s + a_0}{s^n + b_{n-1} s^{n-1} + \cdots + b_1 s + b_0} \tag{1}$$

The roots of the numerator are called the zeros and the roots of the denominator are called the poles. They can be ploted in the s complex plane as circles and crosses. The properties of a transfer function can be completely determined from its poles and zeros. For example, for stability reasons the poles must lie in the left-half s-plane.

A special case in the analysis of the properties of a transfer function is the case when the input signal is a sine wave. In this case $s = j\omega$ and the transfer function becomes

$$N(j\omega) = H \frac{a_n (j\omega)^n + a_{n-1} (j\omega)^{n-1} + \cdots + a_1 j\omega + a_0}{(j\omega)^n + b_{n-1} (j\omega)^{n-1} + \cdots + b_1 j\omega + b_0} \tag{2}$$

which can be written as

Approximation in Analog Signal Processing

$$N(j\omega) = \frac{a_0 - a_2\omega^2 + a_4\omega^4 - \cdots + j(a_1\omega - a_3\omega^3 + \cdots)}{b_0 - b_2\omega^2 + b_4\omega^4 - \cdots + j(b_1\omega - b_3\omega^3 + \cdots)} \qquad (3)$$

Its magnitude squared is then given by

$$|N(j\omega)|^2 = \frac{(a_0 - a_2\omega^2 + a_4\omega^4 - \cdots)^2 + (a_1\omega - a_3\omega^3 + \cdots)^2}{(b_0 - b_2\omega^2 + b_4\omega^4 - \cdots)^2 + (b_1\omega - b_3\omega^3 + \cdots)^2} \qquad (4)$$

After developing the parentheses, we get only even powers of ω in the following way

$$|N(j\omega)|^2 = H^2 \frac{c_0 + c_1\omega^2 + c_2\omega^4 + c_3\omega^6 + \cdots}{d_0 + d_1\omega^2 + d_2\omega^4 + d_3\omega^6 + \cdots} \qquad (5)$$

Thus, a m agnitude sq uared has on ly ev en p owers of ω and it is, therefore, an even function of ω. On the other hand, the phase or argument of Eq. 3 is given by

$$\text{Arg } N(j\omega) = \tan^{-1}\left(\frac{a_1\omega - a_3\omega^3 + \cdots}{a_0 - a_2\omega^2 + a_4\omega^4 - \cdots}\right) - \tan^{-1}\left(\frac{b_1\omega - b_3\omega^3 + \cdots}{b_0 - b_2\omega^2 + b_4\omega^4 - \cdots}\right) \qquad (6)$$

Since the inverse tangent is an odd function of ω, the phase is, therefore, an odd function of ω. We will use th e pro perties d eveloped i n th is sect ion in the re maining s ections of t he chapter. I n the foll owing s ections we w ill be m ainly conc erned wit h low p ass fun ctions. A low pass function may have all of its zeros at infinity, in this case the function has the form

$$N_{pb}(s) = \frac{H}{s^n + a_n s^{n-1} + \cdots + a_1 s + a_0} \qquad (3)$$

A magnitude plot for t his f unction has the form show n in Fig ure 1. In t his fi gure w e define the band pass from $\omega = 0$ rad/sec to ω_c, and the stop band from ω_s to infinity. The band between ω_c and ω_s is called the transition b and. In this b and the m agnitude ch aracteristic is usually stee p. We can see tolera nce bands in t he p ass ban d an d at the st opband. The tolerances are gi ven i n terms of atte nuation in dB. In th e p ass band t his is t he maximum tolerance al lowed, th us i s is a bbreviated A_{max}. On the other hand, the tolerance in the st op band is a minimum atten uation. Th is means that we coul d ac hieve more thatn th at req uired but n o less attenuation. T herefore, i t is called the m inimum atten uation and is abbreviated A_{min}.

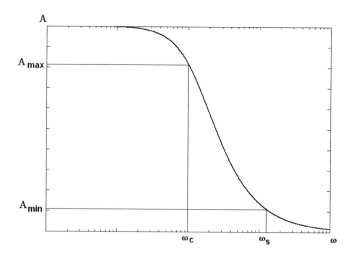

Figure 1. Low pass function parameters.

Maximally Flat Magnitude Approximation

To develop the concept of maximally flat magnitude, let us consider Eq. 5. We rewrite it as

$$|N(j\omega)|^2 = H_0^2 \frac{1+\alpha_1\omega^2+\alpha_2\omega^4+\alpha_3\omega^6+....}{1+\beta_1\omega^2+\beta_2\omega^4+\beta_3\omega^6+...} \qquad (7)$$

The quotient indicated can be divided to become to

$$|N(j\omega)|^2 = H_0^2\left[1+(\alpha_1-\beta_1)\omega^2+(\alpha_2-\beta_2+\beta_1^2-\beta_1\alpha_1)\omega^4+...\right] \qquad (8)$$

This is a power series in ω. That is, it is a Maclaurin series around the origen. The coefficients of the series are the derivatives of the magnitude squared. We note that there are no odd powers of ω. Thus, the odd order derivatives are equal to zero already. To make the magnitude as flat as possible we need as many derivatives possible equal to zero. Thus we need to make as many coefficients in Eq. 8 equal to zero. Then, from Eq. 8, by making $\alpha_1-\beta_1=0$ we make the second derivative equal to zero. Because we do not have the third power of w, we have a third derivative equal to zero. We can make the fourth derivative zero by making $\alpha_2-\beta_2=0$. In general, we can make as many derivatives equal to zero as we can by simply satisfying:

$$\alpha_i = \beta_i \qquad (9)$$

For as many coefficients as we can. Due to the flatness at the origin, this kind of filters is known as maximally flat magnitude filters, or MFM. An example shows the procedure.

Approximation in Analog Signal Processing

Example 1. It is desired to find the coefficient values for the function $N(s)$ to have a maximally flat magnitude, where $N(s)$ is given by

$$T(s) = \frac{s+z}{s^2 + as + b}$$

First we need to make $s = j\omega$ to obtain $|T(j\omega)|^2$ which can be written as the equation in the form of Eq. 7. The, we get

$$|T(j\omega)|^2 = \frac{z^2}{b^2} \frac{1 + (1/z^2)\,\omega^2}{1 + [(a^2 - 2b)/b^2]\,\omega^2 + (1/b^2)\,\omega^4}$$

Equating the coefficients of ω^2 in the numerator and denominator, we find that the function has a maximally flat magnitude if the coefficients satisfy the condition

$$a^2 - 2b = \left(\frac{b}{z}\right)^2$$

Butterworth Approximation

A Butterworth approximation has a maximally flat magnitude at the origin, that is at DC. We assume that the zeros of the transfer function are at infinity, thus, the transfer function magnitude squared has the form

$$|N(j\omega)|^2 = H^2 \frac{1}{1 + a_1\omega^2 + a_2\omega^4 + a_3\omega^6 + \ldots} \tag{10}$$

Using Eq. 9 we see that the coefficients of the denominator polynomial must be equal to zero (because the numerator coefficients are zero) except for the last one. Thus, we have that the resulting transfer function is

$$|N(j\omega)|^2 = \frac{H^2}{1 + a_n\omega^{2n}} \tag{11}$$

The coefficient a_n is ussually written as ε^2. Thus, Eq. 11 changes to

$$|N(j\omega)|^2 = \frac{H^2}{1 + \varepsilon^2\omega^{2n}} \tag{12}$$

This function is called a *Butterworth function*. The variable ε is related to A_{max} by the following equation

$$\varepsilon = \sqrt{10^{A_{max}/10} - 1} \tag{13}$$

To ob tain th e Bu tterworth transfer fun ction $N(s)$ w e make a c hange of variable $\omega = s/j$ to obtain

$$N(s)N(-s) = \frac{H^2}{1 + \varepsilon^2 \left(\dfrac{s}{j}\right)^{2n}} \tag{14}$$

To obtain the poles we equate the d enominator of this e quation to zero. It can be sh own that the poles are given by

$$p_k = -\frac{1}{\sqrt[n]{\varepsilon}} \sin\frac{2k-1}{2n}\pi + j\frac{1}{\sqrt[n]{\varepsilon}}\cos\frac{2k-1}{2n}\pi \qquad k = 1,\ 2,\ 3,...,\ n \tag{15}$$

Where w e h ave c hosen o nly the lef t-hand pl ane p oles for stabi lity reasons. A plot of two examples is show n in Figur e 2. T here we see that B utterworth poles are equally-space d and that they lie on a circle of radius $1/\sqrt[n]{\varepsilon}$. In addition, w e see that for odd order cases there is a real pole whereas that even order cases only have complex conjugate poles.

The order of a Butterworth function can be determined from

$$n \geq \frac{\log\left[\dfrac{10^{A_{min}/10} - 1}{\varepsilon^2}\right]^2}{\log(\omega_s/\omega_c)} = \frac{\log\left[\dfrac{10^{A_{min}/10} - 1}{10^{A_{max}/10} - 1}\right]^2}{\log(\omega_s/\omega_c)} \tag{16}$$

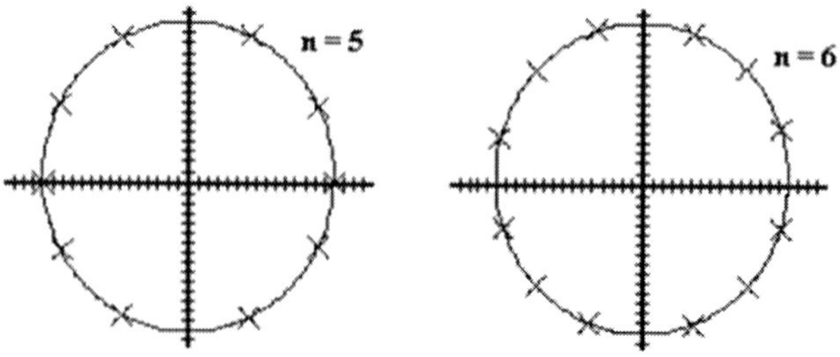

Figure 2. Butterworth poles for $N(s)N(-s)$. Note the equal spacing between poles.

The greater than or equal sign is used because the order must be an integer number. Thus, the order is the integer grater that the result given by Eq. 16.

Example 2. We wish to determine the order of a normalized Butterworth filter that must satisfy a passband tolerance of 2 dB, a stopband attenuation of at least 30 dB at 2 rad/s. The data is the $A_{max} = 2$, $A_{min} = 30$, $\omega_c = 1$ and $\omega_s = 2$.

Using Eq. 15 we have

$$n \geq = \frac{\log\left[\dfrac{10^{30/10}-1}{10^{2/10}-1}\right]^2}{\log(2/1)} = 5.37$$

Thus the required order is 6.

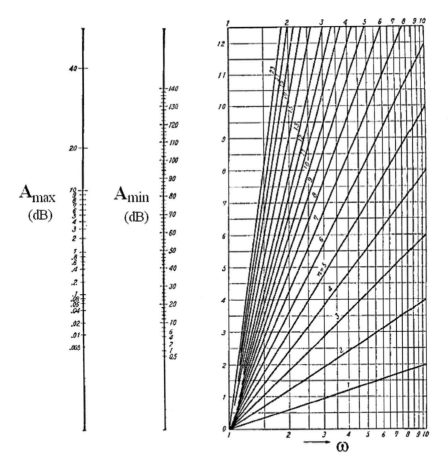

Figure 3a. Nomograph for Butterworth filters.

An alternative way to calculate the order is by means of the Kawakami nomographs. The nomograph for Butterworth filters is shown in Figure 3a. Its use is illustrated in Figure 3b. For Example 2, we can readily see that the order lies between the lines corresponding to $n = 5$ and $n = 6$. Thus, the order is $n = 6$, as shown in Figure 4.

Figure 5 shows plots of the magnitude of Butterworth functions for several orders. We can see that as the order increases the roll off at the band edge is steeper. The price we pay for an increase in the order is a circuit with more elements and, therefor, a more costly circuit.

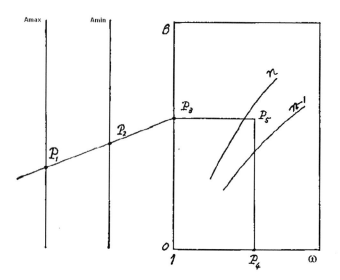

Figure 3b. Use of the nomograph.

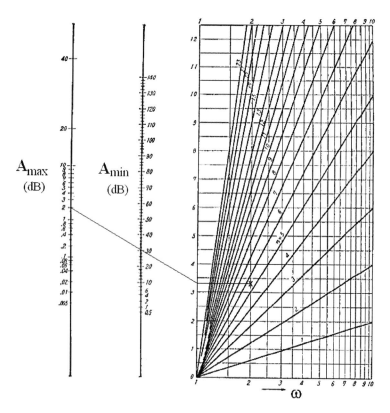

Figure 4. Use of the nomograph for Example 2.

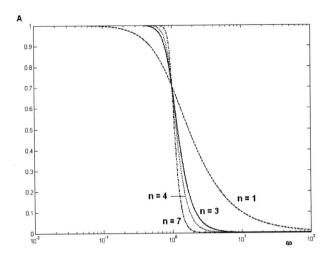

Figure 5. Magnitude plots for Butterworth filters.

Chebyshev Approximation

The Chebyshev approximation uses Chebyshev polynomial for the introduction of an equal ripple at the passband. Figure 6 shows plots of Chebyshev filters of sixth and seventh order. We can clearly appreciate the ripple. The complexity of the pole equations does not increase in mathematical complexity but we gain in a steeper roll off at the band edge, and thus, a reduced order, and a cheaper circuit realization.

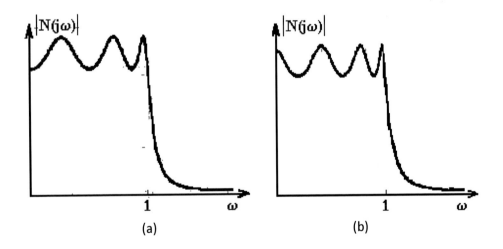

Figure 6. Magnitude plots for Chebyshev filters.

The magnitude squared Chebyshev approximation is given by

$$|N(j\omega)|^2 = \frac{H^2}{1+\varepsilon^2 C_n^2(\omega)} \tag{17}$$

Where $C_n(\omega)$ is a Chebyshev polynomial. The first few Chebyshev polynomials are ploted in Figure 7 and are given by

$$C_1(\omega) = \omega$$
$$C_2(\omega) = 2\omega^2 - 1$$
$$C_3(\omega) = 4\omega^3 - 3\omega$$
$$C_4(\omega) = 8\omega^4 - 8\omega^2 + 1$$
$$\vdots$$
$$C_{n+1}(\omega) = 2\omega\, C_n(\omega) - C_{n-1}(\omega)$$
(18)

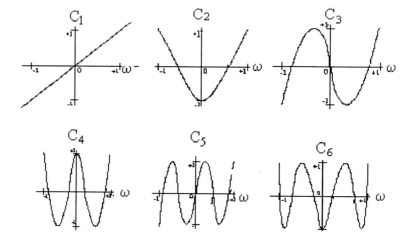

Figure 7. The first six Chebyshev polynomials.

The poles for Chebyshev filters can be found from

$$p_k = -\sigma_k + j\omega_k = -\sin u_k \sinh v + j \cos u_k \cosh v \qquad (19)$$

where

$$u_k = \frac{2k-1}{2n}\pi \qquad k = 1, 2, \ldots, 2n \qquad (20)$$

and

$$v = \frac{1}{n}\sinh^{-1}\frac{1}{\varepsilon} \qquad (21)$$

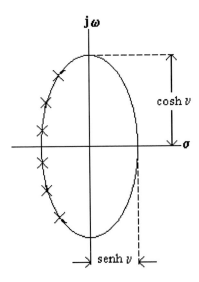

Figure 8. Locus for Chebyshev poles.

Chebyshev poles are located on an ellipse in the complex plane. If σ_k and ω_k are the real and imaginary parts, its location is given by the equation

$$\frac{\sigma_k^2}{\sinh^2 v} + \frac{\omega_k^2}{\cosh^2 v} = 1$$

and shown in Figure 8. The order n can be calculated from

$$n \geq \frac{\cosh^{-1} \sqrt{\dfrac{10^{A_{min}/10} - 1}{\varepsilon^2}}}{\cosh^{-1}\left(\dfrac{\omega_s}{\omega_c}\right)} \tag{22}$$

As with Butterworth filters, there is a nomograph for order determination. It is shown in Figure 9.

Example 3 It is desired to find the poles of a Chebyshev function with a ripple of 1 dB in the pass band that goes from dc to 1 r/s, and with $n = 5$. Using Eq. 13 we have that

$$\varepsilon = \sqrt{10^{A_{max}/10} - 1} = \sqrt{10^{1/10} - 1} = 0.508847139909588$$

We now obtain v from Eq. 21 as

$$v = \frac{1}{n}\sinh^{-1}\frac{1}{\varepsilon} = \frac{1}{5}\sinh^{-1}\frac{1}{0.508847139909588} = 0.285595071772705$$

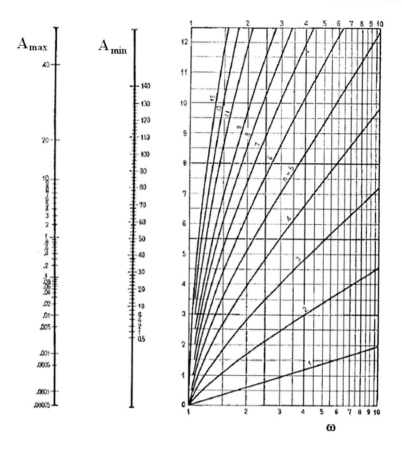

Figure 9. Nomograph for Chebyshev filters.

And u_k from Eq. 20

$$u_1 = \frac{2-1}{2 \times 5}\pi = \frac{\pi}{10}$$

$$u_2 = \frac{2 \times 2 - 1}{2 \times 5}\pi = \frac{3\pi}{10}$$

$$u_3 = \frac{2 \times 3 - 1}{2 \times 5}\pi = \frac{5\pi}{10}$$

$$u_4 = \frac{2 \times 4 - 1}{2 \times 5}\pi = \frac{7\pi}{10}$$

$$u_5 = \frac{2 \times 5 - 1}{2 \times 5}\pi = \frac{9\pi}{10}$$

Finally, the poles are evaluated from Eq. 19 and we get

$$k = 1 \quad \sigma_1 = -\sin\frac{\pi}{10}\sinh v = -0.08946 \qquad \omega_1 = \cos\frac{\pi}{10}\cosh v = 0.99011$$

$$k = 2 \quad \sigma_2 = -\sin\frac{3\pi}{10}\sinh v = -0.23421 \qquad \omega_2 = \cos\frac{3\pi}{10}\cosh v = 0.61192$$

$$k = 3 \quad \sigma_3 = -\sin\frac{5\pi}{10}\sinh v = -0.28949 \qquad \omega_3 = \cos\frac{5\pi}{10}\cosh v = 0$$

$$k = 4 \quad \sigma_4 = -\sin\frac{7\pi}{10}\sinh v = -0.23421 \qquad \omega_4 = \cos\frac{7\pi}{10}\cosh v = 0.61192$$

$$k = 5 \quad \sigma_5 = -\sin\frac{9\pi}{10}\sinh v = -0.08946 \qquad \omega_5 = \cos\frac{9\pi}{10}\cosh v = 0.99011$$

Inverse Chebyshev Approximation

An inverse-Chebyshev approximation has a monotonic response in the passband and an equal ripple in the stopban, as opposed to the regular Chebyshev characteristic. In digital signal processing the Chebyshev characteristic is known as Chebyshev I and the inverse-Chebyshev characteristic is known as Chebyshev II.

The magnitude squared function for an inverse-Chebyshev function is

$$\left| N_{CI}(j\omega) \right|^2 = \frac{\varepsilon^2 C_n^2\left(\frac{1}{\omega}\right)}{1 + \varepsilon^2 C_n^2\left(\frac{1}{\omega}\right)} \tag{23}$$

And a plot of the magnitude response is shown in Figure 10. Note that now, the parameter e is associated with the stopband ripple and that the normalized frequency is the stopband frequency. The poles of the inverse-Chebyshev characteristic are given by:

$$p_k = \frac{1}{\sigma_k + j\omega_k} \qquad k = 1, 2, \ldots, n \tag{24}$$

Where the parameters σ_k and ω_k are given in Eq. 18, but the value of e is changed to

$$\varepsilon^2 = \frac{1}{10^{0.1 A_{min}} - 1} \tag{25}$$

The stopband zeros, which are located on the $j\omega$ axis are given by

$$z_k = j\dfrac{1}{\cos\left(\dfrac{2k-1}{2n}\pi\right)} \qquad k = 1,2,\ldots,n \tag{26}$$

The or der ca n be calculated w ith th e sam e equat ion us ed for Che byshev functions. Furthermore, the sam e nom ograph c an be use d. W e only n eed to exchange passband a nd stopband frequencies. We may ask w hat is the advantage of using inverse-Chabyshev filters. The answ er is a better tim e delay as compared to Che byshev fi lters. Th is be havior is examined in the phase approximation section.

Example 4 We wish to obtain the order of an i nverse-Chebyshev function with a pass b and ripple of 0.1 dB in the band from dc to 1 rad/sec, and a stopband attenuation of at least 30 dB at frequencies greater that 2 rad/sec.

Since $A_{\max} = 0.1$ dB, $A_{\min} = 30$dB, $\omega_c = 1$ rad/sec and $\omega_s = 2$ rad/sec, we have

$$n \geq \dfrac{\cosh^{-1}\left(\dfrac{10^3 - 1}{10^{0.01} - 1}\right)^{1/2}}{\cosh^{-1}\left(\dfrac{2}{1}\right)} = 4.5759$$

Thus, the required order is 5.

Elliptic Approximation

An elliptic tra nsfer function is called this way because it makes use of Jacobian elliptic functions. The elliptic filter transfer functions were developed by Wilhelm Caur en the 1930's and they have pro ved to be the best in terms of magnitude c haracteristic as compared to the previous characteristics studied so far.

An elliptic magnitude squared function has the form

$$\left|N(j\omega)\right|^2 = \dfrac{H^2}{1 + \varepsilon^2 R_n^2(\omega)} \tag{27}$$

where $R_n(\omega)$ is a rational function of ω. The usual form for this rational function is

$$R_n(\omega) = M\prod_{i=1}^{\frac{n}{2}} \dfrac{\omega^2 - \Omega_{Pi}^2}{\omega^2 - \Omega_{Ci}^2} \qquad n \quad \text{even} \tag{28a}$$

$$R_n(\omega) = M\omega\prod_{i=1}^{\frac{n-1}{2}} \dfrac{\omega^2 - \Omega_{Pi}^2}{\omega^2 - \Omega_{Ci}^2} \qquad n \quad \text{odd} \tag{28b}$$

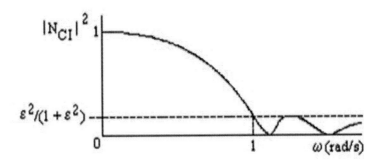

Figure 10. Inverse-Chebyshev magnitude characteristic.

Where the parameters Ω_k are the locations of maxima and minima in the passband and stopband. A typical low pass elliptic characteristic is shown in Figure 11. The zeros in the stopband are a consecuence of the rational functions from Eq. 27. Typical transfer function for elliptic filters are

$$N_I(s) = H_0 \frac{\prod_{i=1}^{\frac{n-1}{2}}\left(s^2 + \Omega_{Ci}^2\right)}{a_0 + a_1 s + a_2 s^2 + \ldots + s^n} \qquad \text{odd order} \qquad (29a)$$

$$N_p(s) = H_0 \frac{\prod_{i=1}^{\frac{n}{2}}\left(s^2 + \Omega_{ai}^2\right)}{a_0 + a_1 s + a_2 s^2 + \ldots + s^n} \qquad \text{even order} \qquad (29b)$$

The computation and calculation of poles and zeros for elliptic filters usually is done by computer programs because it requires extensive calculations. The order can be obtained with the use of the Kawakami nomograph shown in Figure 12.

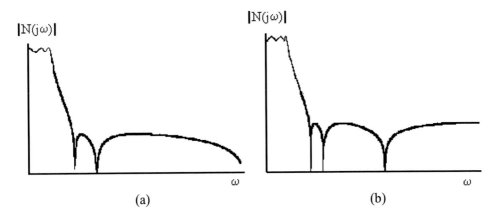

Figure 11. Elliptic filter .a) odd order (fifth order) , (b) even order (sixth order).

Phase Approximation

The ideal phase in any analog filter is a linear phase. The advantage of a linear phase can be seen when we process signals of different frequencies, the filter is going to process them with the same delay. To show this we define the group delay $D(\omega)$ as

$$D(\omega) = -\frac{d}{d\omega}\arg N(j\omega) \qquad (30)$$

If the phase is a linear function

$$\arg N(j\omega) = -\omega\, t_0 \qquad (31)$$

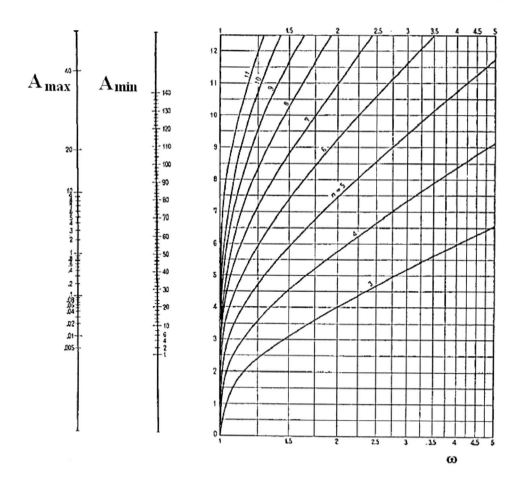

Figure 12. Nomograph for elliptic filter order computation.

Then, the delay is given by

$$D(\omega) = t_0 \qquad (32)$$

Approximation in Analog Signal Processing

Which has a constant value and it is independent of frequency.

In act ual analog fil ters, it is not possible to obtain a line ar phase, but inste ad, we can approximate a linear phase. A way to produce a linear phase is by making Taylor expansion around the origin of the phase function of the form

$$\arg N(j\omega) = a_1\omega + a_2\omega^3 + a_3\omega^5 + \cdots \tag{33}$$

Note th at we do no t ha ve even p owers beca use acc ording to Eq. x the p hase is an od d function. Furtherm ore, it is is not a line ar f unction becaus e of the higher order term s in addition to the linear one. Thus, in order to make the phase as close as possible to a linear one we need to make the coefficients $a_2 = a_3 = a_4 = \ldots = 0$. How many coefficients can made zero depends upon the degrees of freedom the transfer function posses, that is, the number of poles that the transfer function has.

Thomson Approximation

The problem posed in the preceding section was solved by W. E Thomson. He found that the denominator po lynomial for a li near phase approximation w ere gi ven by Bessel polynomials. Bessel polynomials $B(s)$ follow the recusrion formula

$$B_n(s) = (2n-1)B_{n-1}(s) + s^2 B_{n-2}(s) \tag{34}$$

and the first two polynomials are $B_1(s) = s+1$, and $B_2(s) = s^2 + 3s + 3$. The coefficients of Bessel polynomials are given by

$$a_k = \frac{(2n-k)!}{2^{n-k}k!(n-k)!} \qquad k = 0,1,\cdots,n-1 \tag{35}$$

where n is the order. Then, the transfer function that approximates linear phase is given by

$$N(s) = \frac{a_0}{B_n(s)} \tag{36}$$

Since th e denominator p olynomial is a Bessel polynomial, an other name give n to tra nsfer functions with linear phase is Bessel filters and Bessel-Thomson filters.

From Eq. 30, the delay has the form

$$D(\omega) = a_1 + 2a_2\omega + 3a_3\omega^2 + \cdots \tag{37}$$

And since we wish to have a linear phase, then we need the coefficients to satisfy

$$a_1 \neq 0$$
$$a_2 = 0 \qquad (24)$$
$$a_3 = 0$$
$$\vdots$$

A plot with the delay for several filter orders is given in Figure 13. Since the delay in this kind of filters is flat, they are also known as maximally delay, or MFD, filters. To have an idea of the improvement in delay flatness, Figure 14 shows the delay for Butterworth and Chebyshev filters.

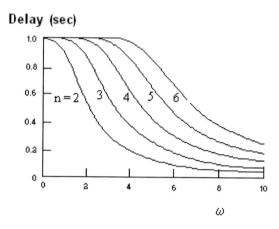

Figure 13. Delay for Thomson filters.

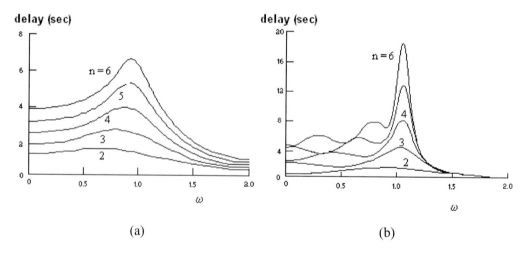

Figure 14. Delay for a) Butterworth and b) Chebyshev filters.

The order can be obtained from delay information: If T is the desired delay at a given frequency w and A(w) is the loss at dB at that frequency the order can be evaluated from

$$n \geq \frac{5(\omega T)^2 \log e}{A(\omega)} + \frac{1}{2} \qquad (25)$$

Fianally, fig ure 16 sh ows Thom son filter po le l ocations together w ith Bu tterworth an d Chebyshev poles. We note that Thomson poles are farther away from the origin than other filter type poles are.

It can be shown that the error in the delay is given by

$$\text{Error in } D(\omega) = \frac{e^{-(\omega t)^2/(2n-1)}(\omega T)^{2n}}{a_0^2} \qquad (36)$$

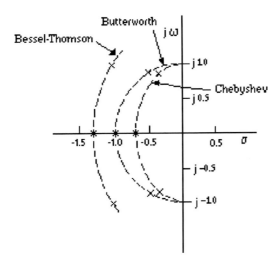

Figure 16. Pole locations for Thomson, Butterworth and Chebyshev low pass filters.

Ejemplo 6 Find the order and the transfer function for a Thomson filter that has a delay of 10 μs at low freque ncies an d an error of 1 % for $f \leq 20$ kH z. The d eviation i n the attenuation must not be greater than 1 dB in the range from dc to 20 kHz. Using Eq. 36:

$$n \geq \frac{5\left[(2\pi \times 20 \times 10^3)(10 \times 10^{-6})\right]^2 \log e}{1} + \frac{1}{2} = 3.92$$

Thus, the required order is $n = 4$. The error in the delay is

$$\text{Error in } D(\omega) = \frac{\left[(2\pi \times 20 \times 10^3)(10 \times 10^{-6})\right]^8 e^{-(\omega T)^2/7}}{105^2}$$

$$= 4.48 \times 10^{-4}$$

Which is much less tan the required 1%. Finally, it can be shown that the transfer function is

$$N(s) = \frac{105}{s^4 + 10s^3 + 45s^2 + 105s + 105}$$

Frequency Transformations

So far in the chapter we have been concerned with low pass transfer function. In this section we see how to transform them to high pass, band pass and band reject filters.

To convert a low pass filter to a high pass one we make the change of variable

$$s \to \frac{1}{s} \tag{26}$$

This moves the poles from the original locations to new location such that if they were inside the unit circle, they will move to outside the unit circle and viceversa. The transfer function now has as many zeros at the origin as poles in it. A typical high pass Butterworth, Chebyshev, and Thomson function is of the form

$$N_{HP}(s) = H \frac{s^n}{s^n + b_{n-1}s^{n-1} + \cdots + b_1 s + b_0} \tag{27}$$

As a rule, the degrees of the numerator and denominator polynomials are equal. Because of the form of the numerator, there exist n zeros at the origin giving place to the high pass behavior.

To convert a low pass function to a band pass function we first define in Figure 15 the parameters of a band pass function

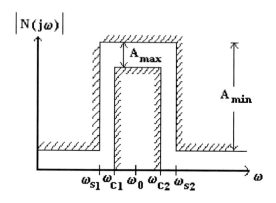

Figure 15. Parameters of a band pass function.

Where the frecuencies that define the band pass B given by ω_{p1} and ω_{p2}. Two important parameters are the center frequency ω_0 and the quality factor Q defined by

$$\omega_0^2 = \omega_{c\,1}\omega_{c\,2} \tag{28}$$

$$B = \omega_{c2} - \omega_{c1} \tag{29}$$

$$Q = \frac{\omega_0}{B} \tag{30}$$

To obtain the band pass we use the so-called low pass to band pass transformation

$$s \rightarrow \frac{1}{B}\left(s + \frac{\omega_o^2}{s}\right) \tag{31}$$

Applying this transform ation to a low pass fu nction w e o btain a ba nd pass tr ansfer function which is of the form

$$N_{BP}(s) = \frac{Hs^n}{s^{2n} + b_{2n-1}s^{2n-1} + \cdots + b_2 s^2 + b_1 s + b_0} \tag{32}$$

Note th at th e order of a ban d pass tra nsfer function num erator is ha lf t he de nominator polynomial order. This im plies that there are n zeros at the origin and n zeros at infinity and $2n$ poles producing the band pass behavior.

Finally, to obtain a band reject transformation we use the

$$s \rightarrow \frac{pB}{s^2 + \omega_o^2} \tag{33}$$

A band reject function is of the form

$$N_{RB}(p) = \frac{H\left(p^2 + \omega_o^2\right)^n}{p^{2n} + a_{2n-1}p^{2n-1} + a_{2n-2}p^{2n-2} + \cdots + a_1 p + 1} \tag{34}$$

Note that if we start with an nth order lowpass function we end up with a band reject function that has $2n$ poles and $2n$ complex conjugate zeros on the im aginary axis, n zeros at $j\omega_o$ and n zeros at $-j\omega_o$.

Concluding Remarks

We ha ve covered the t opic of approximation in an alog fi lter design. Th e fu nctions covered w ere Bu tterworth, Ch ebyshev, i nverse-Chebyshev, e lliptic, an d T homson. The las t one is a linear phase ap proximation a nd the rem aining on es are m agnitude approximations. All of thes e a pproximations are for low pass func tions. The las t se ction c overs freque ncy transformations that allows us t o obtain high pass, band pass, and band reject functions from low pass prototype functions.

References

L. P. H uelsman and P.E. Allen, (2 008). Intro duction to the T heory an d D esign of A ctive Filters, McGraw-Hill Book Co., N.Y., 1980.

Thomson, W.E., "Delay Networks having Ma ximally Flat Frequency Characteristic s", *Proceedings of the Institution of Electrical Engineers*, Part III, Novem ber 1949, Vol. 96, No. 44, pp. 487-490.

S. Butterworth, *"On the Theory of Filter Amplifiers"*, *Wireless Engineer*, vol. 7, pp. 536-541, 1930.

W. Cauer. *Theorie der linearen Wechselstromschaltungen*, Vol. I. Akad. Verlags-Gesellschaft Becker und Erler, Leipzig, 1941.

R. Sch aumann, X . H aiqiao, an d M . V anValkenburg, *Design of Analog Filters*, 2n d Edition, Oxford University Press, 2009.

R. Sch aumann, X . H aiqiao, an d M . V anValkenburg, *Design of Analog Filters*, 2n d Edition, Oxford University Press, 2009.

PART II: APPLICATIONS

In: Analog Circuits: Applications, Design and Performance
Editor: Esteban Tlelo-Cuautle

ISBN: 978-1-61324-355-8
© 2012 Nova Science Publishers, Inc.

Chapter 4

TRANSCONDUCTANCE AMPLIFIERS: NAM REALIZATIONS AND APPLICATIONS

Ahmed M. Soliman[*]

Electronics and Communication Engineering Department
Faculty of Engineering Cairo University, Giza 12613, Egypt

Abstract

A systematic generation method of the single input single output transconductance amplifier (TA) based on using nodal admittance matrix (NAM) expansion is given. The four pathological elements used are the nullator, norator, voltage mirror (VM) and current mirror (CM). The single input single output TA also known as the voltage controlled current source (VCCS) includes two types depending on the direction of the output current. Two pathological realizations for each type using grounded resistor are given and eight pathological realizations for each type using floating resistor are also derived. Applications of TA in realizing grounded and floating resistors, grounded and floating inductors, first order voltage mode and current mode all pass filters, Tow Thomas second order filter, universal second order voltage mode and mixed mode filter using five single input differential output TA and one single input single output TA and oscillators using a single input single output TA are included.

Introduction

This chapter is devoted to the implementation of pathological realizations of the single input single output TA. The use of TA as a basic building block in active circuits has been demonstrated in several papers [1-13]. Only very few nullor representations of TA have appeared in the literature [12-16].

In this chapter the NAM expansion method introduced in [14] to realize controlled sources is extended to accommodate the pathological VM and CM together with the nullator and norator [17] thus resulting in complete set of active circuits realizing the VCCS. For a physically realizable circuit, all the voltages and currents are always uniquely and definitely

---

[*] E-mail address: asoliman@ieee.org

determined. This in turn implies that in the ideal representation of a physically realizable circuit, nullators (or VM) and norators (or CM) must occur in a pair [18-23]. Systematic generation method of controlled sources using unity gain cells has been introduced in the literature [24].

Single Input Single Output TA

There are two types of the single input single output TA according to the direction of the output current.

1. Single Input Single Output TA-

The single input single output TA with output current pointing inwards is shown symbolically in Figure 1(a) and is defined as TA-.

1.1. Pathological Realizations Using a Grounded G

The admittance matrix Y for the TA- is given by:

$$Y = \begin{bmatrix} 0 & 0 \\ G & 0 \end{bmatrix} \tag{1}$$

Adding a third blank row and column and connect a nullator between columns 1 and 3 and a norator between rows 2 and 3in order to move G to the diagonal position 3, 3 therefore the NAM is given by:

$$Y = \begin{bmatrix} 0 & 0 & 0 \\ 0 & 0 & 0 \\ 0 & 0 & G \end{bmatrix} \tag{2}$$

The nullor realization of the above equation is shown in Figure 1(b). An alternative pathological realization can be obtained from eq. (1) by adding a third blank row and column and connect a VM between columns 1 and 3 and a CM between rows 2 and 3in order to move G to the diagonal position 3, 3 therefore the NAM is given by:

$$Y = \begin{bmatrix} 0 & 0 & 0 \\ 0 & 0 & 0 \\ 0 & 0 & G \end{bmatrix} \tag{3}$$

The pathological realization of the above equation is shown in Figure 1(c).

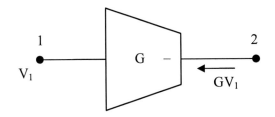

Figure 1(a). Symbolic representation of TA-.

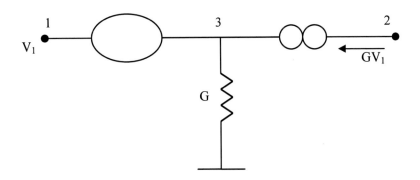

Figure 1(b). Nullator norator realization of TA- [12-16].

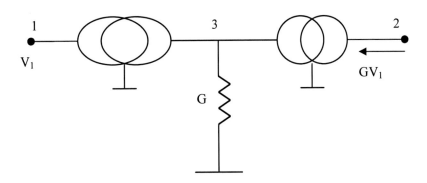

Figure 1(c). New VM-CM realization of TA-.

1.2. Pathological Realizations Using a Floating G

There are eight alternative pathological realizations of the TA- using a floating G as will be explained in this section.

From eq.(1) and adding two blank rows and columns and connect a nullator between columns 1 and 3 and a CM between rows 2 and 4 will move G to the off-diagonal position 4, 3 as -G therefore the NAM is given by:

$$Y = \begin{bmatrix} 0 & 0 & 0 & 0 \\ 0 & 0 & 0 & 0 \\ 0 & 0 & 0 & 0 \\ 0 & 0 & -G & 0 \end{bmatrix} \quad (4)$$

The indefinite admittance matrix has the property that each row and each column sum to zero; hence row zero and column zero contain terms G, - G and G. Connection of a norator between nodes 3 and zero will allow the row zero terms to be brought to row 3. Similarly, a connection of a nullator between nodes 4 and zero will allow the column zero to be brought to column 4 [14] as follows:

$$Y = \begin{bmatrix} 0 & 0 & 0 & 0 \\ 0 & 0 & 0 & 0 \\ 0 & 0 & G & -G \\ 0 & 0 & -G & G \end{bmatrix} \quad (5)$$

The above NAM is realized as shown in Figure 2(a).

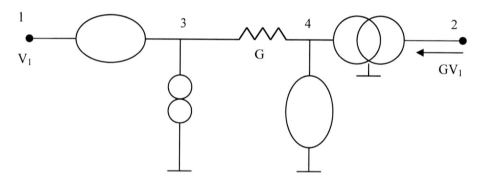

Figure 2(a). Realization 1 of TA- using floating G.

The second realization is obtainable from eq. (1) by adding two blank rows and columns and connect a VM between columns 1 and 3 and a norator between rows 2 and 4 to move G to the off-diagonal position 4, 3 as -G. As in the previous case a connection of a norator between nodes 3 and zero will allow the row zero terms to be brought to row 3 and a connection of a nullator between nodes 4 and zero will allow the column zero to be brought to column 4 [14] as given by the following NAM:

$$Y = \begin{bmatrix} 0 & 0 & 0 & 0 \\ 0 & 0 & 0 & 0 \\ 0 & 0 & G & -G \\ 0 & 0 & -G & G \end{bmatrix} \quad (6)$$

The above NAM is realized as shown in Figure 2(b). The remaining six realizations can be obtained in a similar manner.

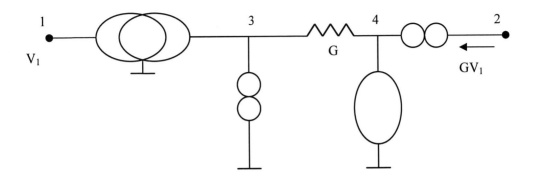

Figure 2(b). Realization 2 of TA- using floating G.

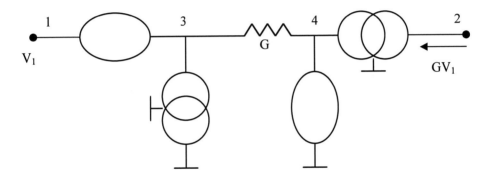

Figure 2(c). Realization 3 of TA- using floating G.

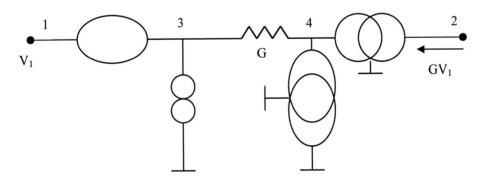

Figure 2(d). Realization 4 of TA- using floating G.

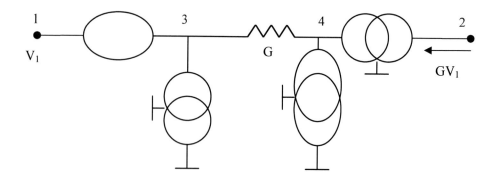

Figure 2(e). Realization 5 of TA- using floating G.

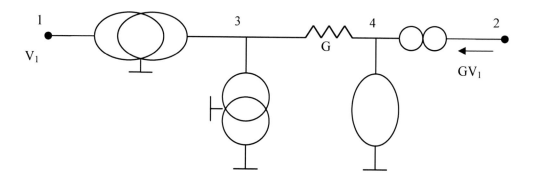

Figure 2(f). Realization 6 of TA- using floating G.

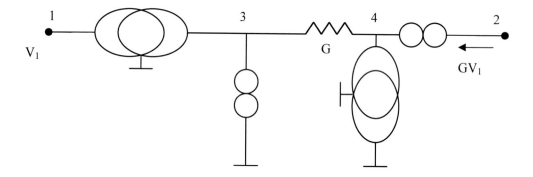

Figure 2(g). Realization 7 of TA- using floating G.

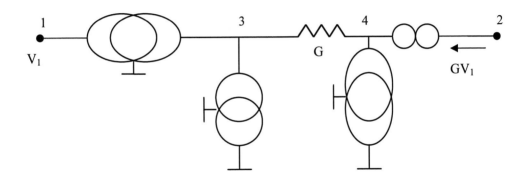

Figure 2(h). Realization of 8 TA- using floating G.

Table I summarizes the properties of the different realizations of the TA-. It is seen that the first two circuits are self adjoint after interchanging ports 1 and 2. Four of the circuits in Figure 2 are adjoints to the other four circuits after interchanging ports 1 and 2 as shown in Table I [25-28].

Table I. Properties of single input single output TA-

Fig. No.	Nullators	VM	Norator	CM	Adjoint to Fig. No.	Self Adjoint
1(b)	1	0	1	0	1(b)	Yes
1(c)	0	1	0	1	1(c)	Yes
2(a)	2	0	1	1	2(b)	No
2(b)	1	1	2	0	2(a)	No
2(c)	2	0	0	2	2(g)	No
2(d)	1	1	1	1	2(f)	No
2(e)	1	1	0	2	2(h)	No
2(f)	1	1	1	1	2(d)	No
2(g)	0	2	2	0	2(c)	No
2(h)	0	2	1	1	2(e)	No

2. Single Input Single Output TA+

The single input single output transconductor with output current pointing outwards is shown symbolically in Figure 3(a) and is defined as TA+.

2.1. Pathological Realizations Using a Grounded G

The admittance matrix Y for the TA+ is given by:

$$Y = \begin{bmatrix} 0 & 0 \\ -G & 0 \end{bmatrix} \quad (7)$$

Adding a third blank row and column and connect a nullator between columns 1 and 3 and a CM between rows 2 and 3 in order to move -G to the diagonal position 3, 3 as +G therefore the NAM is given by:

$$Y = \begin{bmatrix} 0 & 0 & 0 \\ 0 & 0 & 0 \\ 0 & 0 & G \end{bmatrix} \quad (8)$$

The pathological realization of the above equation is shown in Fig. 3(b).

An alternative pathological realization can be obtained from eq. (7) by adding a third blank row and column and connect a VM between columns 1 and 3 and a norator between rows 2 and 3 in order to move -G to the diagonal position 3, 3 as +G therefore the NAM is given by:

$$Y = \begin{bmatrix} 0 & 0 & 0 \\ 0 & 0 & 0 \\ 0 & 0 & G \end{bmatrix} \quad (9)$$

The pathological realization of the above equation is shown in Figure 3(c).

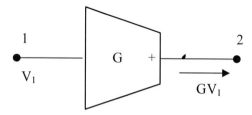

Figure 3(a). Symbolic representation of TA+.

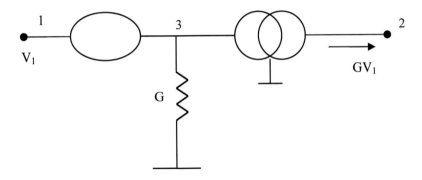

Figure 3(b). New nullator CM realization of TA+.

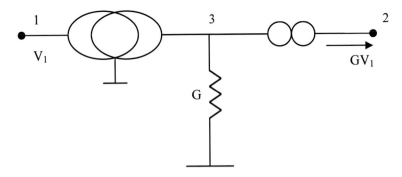

Figure 3(c). New VM norator realization of TA+.

2.2. Pathological Realizations Using a Floating G

There are eight alternative pathological realizations of the TA+ using a floating G as will be explained in this section.
From eq. (7) and adding two blank rows and columns and connect a nullator between columns 1 and 3 and a norator between rows 2 and 4 will move -G to the off-diagonal position 4, 3 therefore the NAM is given by:

$$Y = \begin{bmatrix} 0 & 0 & 0 & 0 \\ 0 & 0 & 0 & 0 \\ 0 & 0 & 0 & 0 \\ 0 & 0 & -G & 0 \end{bmatrix} \qquad (10)$$

Following similar steps as in the previous case the following NAM is obtained.

$$Y = \begin{bmatrix} 0 & 0 & 0 & 0 \\ 0 & 0 & 0 & 0 \\ 0 & 0 & G & -G \\ 0 & 0 & -G & G \end{bmatrix}$$ (11)

The above NAM is realized as shown in Fig.ure 4(a).

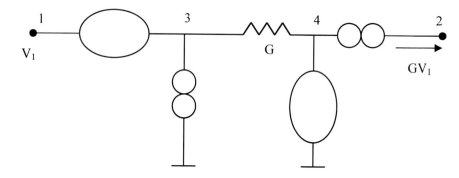

Figure 4(a). Realization 1 of TA+ using floating G [14-16].

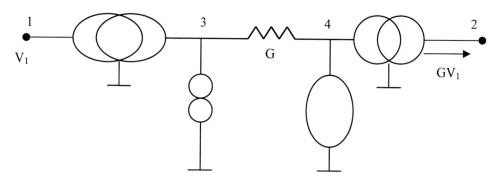

Figure 4(b). Realization 2 of TA+ using floating G.

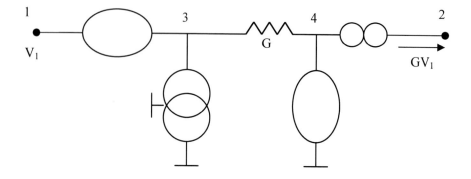

Figure 4(c). Realization 3 of TA+ using floating G.

The remaining seven realizations shown in Figures 4(b) to 4(h) can be obtained in a similar manner. Table II summarizes the properties of the different realizations of the TA+. It is seen that the circuit of Figure 3(b) is adjoint to the circuit of Figure 3(c) after interchanging ports 1 and 2. It is also seen that the circuit of Figures 4(a), (b), (e) and (f) are self adjoint after interchanging ports 1 and 2.

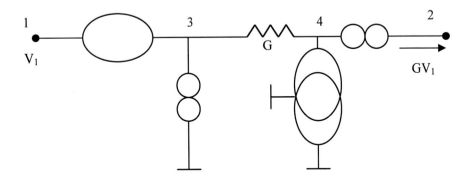

Figure 4(d). Realization 4 of TA+ using floating G.

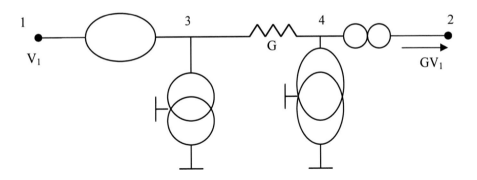

Figure 4(e). Realization 5 of TA+ using floating G.

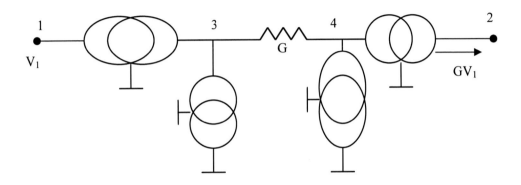

Figure 4(f). Realization 6 of TA+ using floating G.

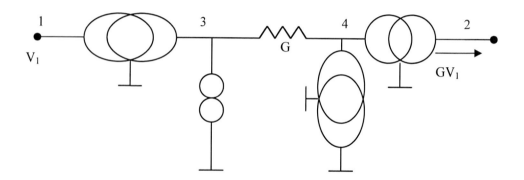

Figure 4(g). Realization 7 of TA+ using floating G.

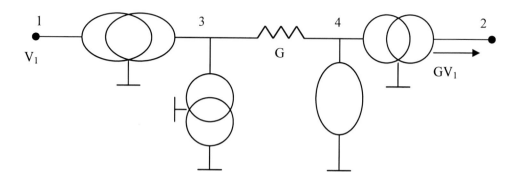

Figure 4(h). Realization 8 of TA+ using floating G.

Table II. Properties of single input single output TA+

Fig. No.	Nullators	VM	Norator	CM	Adjoint to Fig. No	Self Adjoint
3(b)	1	0	0	1	3(c)	No
3(c)	0	1	1	0	3(b)	No
4(a)	2	0	2	0	4(a)	Yes
4(b)	1	1	1	1	4(b)	Yes
4(c)	2	0	1	1	4(d)	No
4(d)	1	1	2	0	4(c)	No
4(e)	1	1	1	1	4(e)	Yes
4(f)	0	2	0	2	4(f)	Yes
4(g)	0	2	1	1	4(h)	No
4(h)	1	1	0	2	4(g)	No

There are several CMOS realizations of TA that are available in the literature [29-31]. Figure 5 represents two CMOS circuits realizing TA- and TA+ that were introduced in [29].

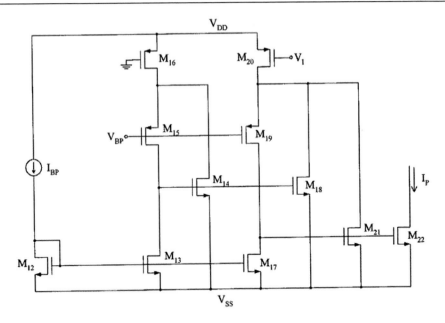

Figure 5(a). CMOS circuit of single input single output TA- [29].

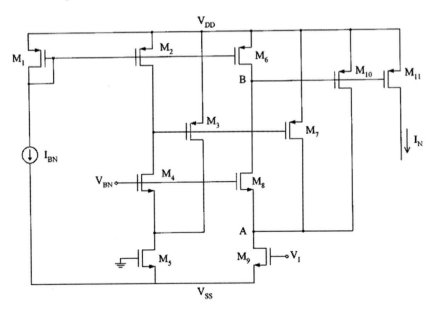

Figure 5(b). CMOS circuit of single input single output TA+ [29].

Table III summarizes different types of TA. The NAM expansion can also be used for all other types of TA resulting in alternative pathological realizations. The pathological realizations are useful in generating practical circuits for different types of TA. As an example consider the pathological realization of the BOTA which is obtained by NAM expansion and shown in Figure 6(a). Figure 6(b) represents a practical BOTA realization using two CCII+.

Table 1II. Symbolic representation of different types of TA

Transconductor	Definition	Symbol
Single input Single output-	$I_2 = GV_1$	
Single input Single output +	$I_2 = -GV_1$	
Single input Balanced output	$I_2 = GV_1$ $I_3 = -GV_1$	
Differential input single output -	$I_3 = G(V_1-V_2)$	
Differential input single output +	$I_3 = -G(V_1-V_2)$	
Differential input balanced output (BOTA)	$I_3 = G(V_1-V_2)$ $I_4 = -G(V_1-V_2)$	

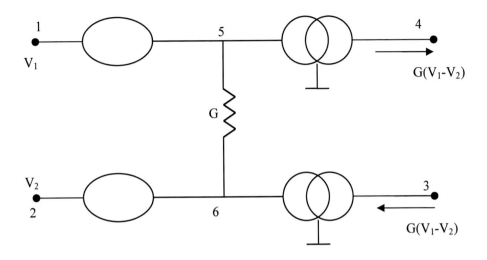

Figure 6(a). Realization of the BOTA using two nullators and two CM.

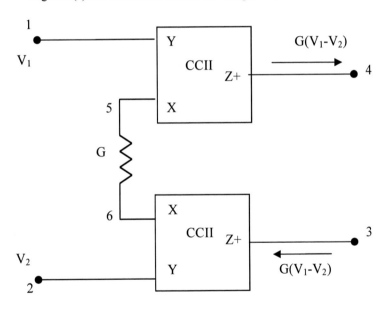

Figure 6(b). Realization of the BOTA using two CCII+.

Applications of TA

Several applications of the TA are available in the literature. This section includes some of the most important applications to demonstrate the importance of TA in active circuits.

1. Realization of Grounded and Floating Resistors from TA

The realization of a grounded resistor using a single input single output TA - is shown in Figure 7(a) [2]. In a similar way a negative resistor can be realized using a single input single output TA + as shown in Figure 7(b).
The single input single output TA published in [29] can be used for this application.

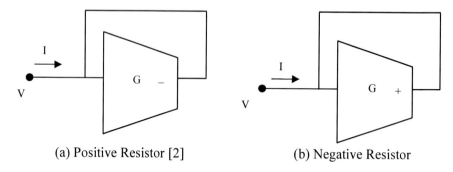

(a) Positive Resistor [2]   (b) Negative Resistor

Figure 7. Realizations of grounded resistor using single TA.

Three alternative realizations for the floating resistor are shown in Figure 8, the first two circuits require that $G_1 = G_2 = G$. The equations describing each of the circuits are given by:

$$I_1 = G(V_1 - V_2), \quad I_2 = -G(V_1 - V_2) \tag{12}$$

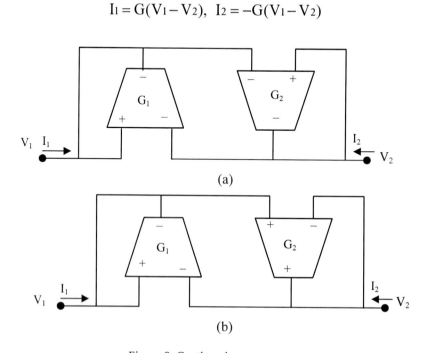

Figure 8. Continued on next page.

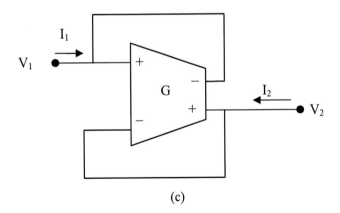

(c)

Figure 8. Three realizations of a floating resistor using TA [2, 10, 32, 33].

2. Realization of Grounded and Floating Inductorsfrom TA

The gyrator circuit shown in Figure 9 realizes a grounded inductor when terminated by a capacitor. The inductance is given by:

$$L = \frac{C}{G_1 G_2} \qquad (13)$$

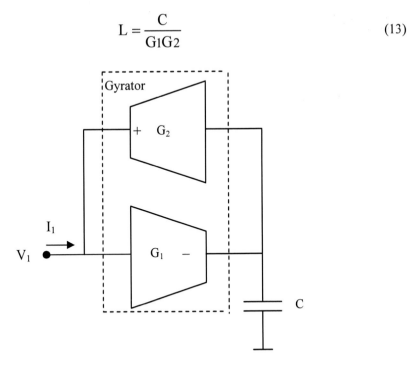

Figure 9. Realization of grounded inductor using two TA.

Two alternative realizations for a floating inductor are shown in Figure 10.
Necessary condition for the circuit of Figures 10(a) is that $G_3 = G_2$ and in this case the circuit equations are given by:

$$I_1 = \frac{G_1G_2}{sC}(V_1-V_2), \quad I_2 = -\frac{G_1G_2}{sC}(V_1-V_2) \tag{14}$$

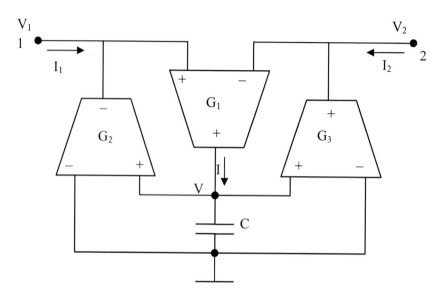

Figure 10(a). Floating L using three TA [32].

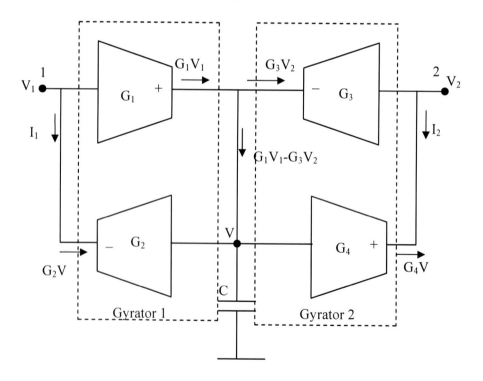

Figure 10(b). Floating L realization using four single input single output TA.

The circuit of Figure 10(b) must satisfy matching conditions $G_3=G_1$ and $G_4=G_2$ and in this case eq.(14) applies to the circuit. This circuit represents the well known realization of a floating inductor from two gyrator circuits with a grounded capacitor inserted between them.

3. Realization of First Order All pass Circuit from TA

Figure 11(a) describes the voltage mode first order all pass [2] having the transfer function:

$$\frac{V_O}{V_I} = \frac{sC - G_1}{sC + \frac{G_1 G_2}{G_3}} \quad (15)$$

A necessary condition for realizing all pass function is that $G_3=G_2$.

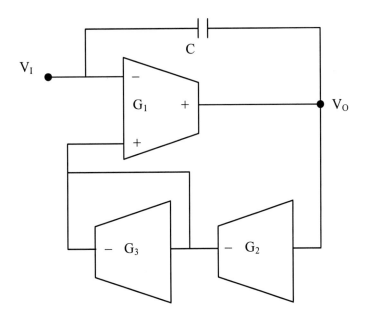

Figure 11(a). Voltage mode first order all pass circuit [2].

Figure 11(b) describes the current mode first order all pass [34] having the transfer function:

$$\frac{I_O}{I_I} = -\frac{sC\frac{G_2}{G_1} - G_2}{sC + G_2} \quad (16)$$

A necessary condition for realizing all pass function is that $G_2=G_1$.

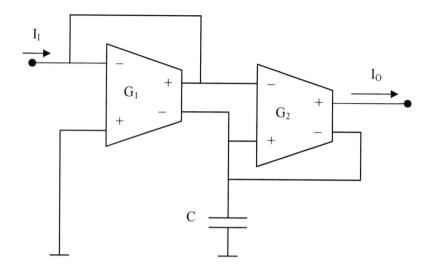

Figure 11(b). Current mode first-order all pass circuit [34].

4. Realization of Tow Thomas Filter Using four TA

The first two integrator loop filter to be considered in this section is the Tow Thomas (TT) filter. The basic building block is the grounded capacitor voltage integrator using a single TA which was originally introduced in [1].
Figure 12(a) represents the inverting TT circuit using four single input single-output TA [4]. The circuit realizes band-pass and low-pass voltage transfer functions given by:

$$\frac{V_{BP}}{V_I} = -\frac{s\ C_2 G_1}{D(s)},\ \frac{V_{LP}}{V_I} = -\frac{G_1 G_4}{D(s)} \tag{17}$$

Where
$$D(s) = s^2 C_1 C_2 + s\ C_2 G_2 + G_3 G_4 \tag{18}$$

The radian frequency and Q of the filter are given by:

$$\omega_o = \sqrt{\frac{G_3 G_4}{C_1 C_2}},\ Q = \frac{1}{G_2}\sqrt{\frac{C_1 G_3 G_4}{C_2}} \tag{19}$$

It is seen that G_2 controls Q without affecting ω_o. It is also seen that G_1 controls the gain without affecting ω_o or Q.

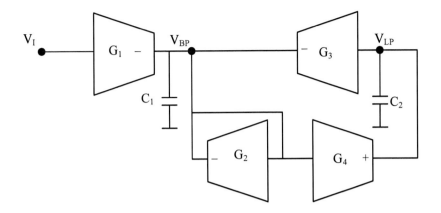

Figure 12(a). Inverting polarity voltage mode TT filter [4].

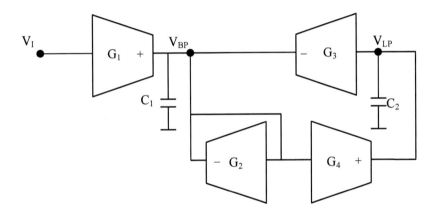

Figure 12(b). Non-inverting polarity voltage mode TT filter [5].

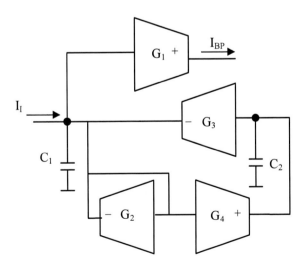

Figure 12(c). Current mode TT band-pass filter [5].

Figure 12(b) represents the non-inverting TT circuit [5] obtained from Figure 13(a) by changing the polarity of G_1 ,eq. (17) apply to this circuit but with positive sign.

The current mode band-pass TT circuit is shown in Figure 12(c) [5] having the same D(s) as given by eq.(18) and its transfer function is given by:

$$\frac{I_{BP}}{I_I} = \frac{s\ C_2 G_1}{D(s)} \tag{20}$$

The current mode low-pass TT filter can be obtained in a similar way.

5. Realization of KHN Filter Using six TA

The second two integrator loop filter considered here is the Kerwin Huelsman Newcomb (KHN) circuit [35]. The KHN circuit using six TA is shown in Figure 13 and is obtained from the KHN circuit using CCII [36].

The transfer functions are given by:

$$\frac{V_{HP}}{V_I} = \frac{s^2 C_1 C_2 G_i}{D(s)}, \quad \frac{V_{BP}}{V_I} = -\frac{s\ C_2 G_1 G_i}{D(s)}, \quad \frac{V_{LP}}{V_I} = \frac{G_1 G_2 G_i}{D(s)} \tag{21}$$

$$\frac{I_{HP1}}{V_I} = \frac{s^2 C_1 C_2 G_1 G_i}{D(s)}, \quad \frac{I_{HP2}}{V_I} = \frac{s^2 C_1 C_2 G_5 G_i}{D(s)} \tag{22}$$

$$\frac{I_{BP1}}{V_I} = -\frac{s\ C_2 G_1 G_2 G_i}{D(s)}, \quad \frac{I_{BP2}}{V_I} = \frac{s\ C_2 G_1 G_4 G_i}{D(s)} \tag{23}$$

$$\frac{I_{LP}}{V_I} = \frac{C_1 G_2 G_3 G_i}{D(s)} \tag{24}$$

Where $$D(s) = s^2 C_1 C_2 G_5 + s\ C_2 G_1 G_4 + G_1 G_2 G_3 \tag{25}$$

The radian frequency and Q of the filter are given by:

$$\omega_0 = \sqrt{\frac{G_1 G_2 G_3}{C_1 C_2 G_5}}\ , Q = \frac{1}{G_4}\sqrt{\frac{C_1 G_2 G_3 G_5}{C_2 G_1}} \tag{26}$$

It is seen that G_4 controls Q without affecting ω_0. It is also seen that G_i controls the gain without affecting ω_0 or Q.

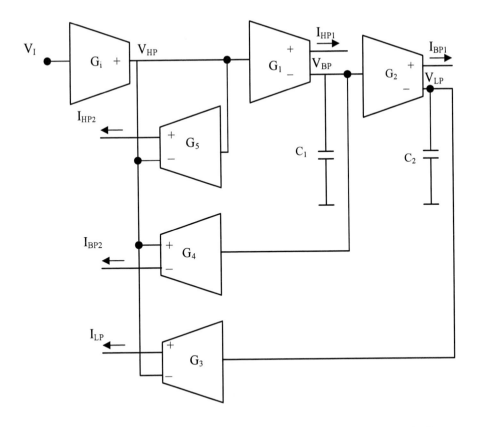

Figure13. KHN voltage mode and mixed mode filter using six TA [35].

6. Realization of Oscillator Using a Single Input Single Output TA

The two oscillators shown in Figure14 were introduced in [37] by systematic generation method using VCCS. The two circuits are adjoint to each other [25-28] thus they have same characteristic equation given by:

$$s^2 C_1 C_2 + s\,[C_2 G_1 + C_1 G_2 + C_2 G_2 - C_2 G_m] + G_1 G_2 = 0 \tag{27}$$

The condition of oscillation and the radian frequency of oscillation are given by:

$$G_m = G_1 + G_2\,[1 + \frac{C_1}{C_2}] \tag{28}$$

$$\omega_o = \sqrt{\frac{G_1 G_2}{C_1 C_2}} \tag{29}$$

It is seen that Gm controls condition of oscillation without affecting radian frequency of oscillation. Most recently the circuit of Figure 14(a) was used as the starting point in the generation of an attractive RF oscillator [38].

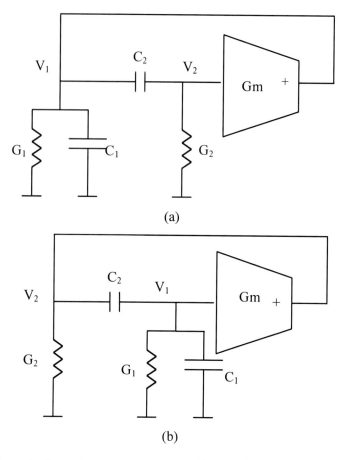

Figure 14. Two minimum component oscillators using a single TA [37].

Conclusion

A systematic generation method of TA based on using nullator, norator elements and pathological mirror elements is used to provide pathological realizations of the single input single output TA. Applications of TA in realizing grounded and floating resistors, grounded and floating inductors, first order all pass filter, universal second order voltage mode and mixed mode filter using five single input differential output TA and one single input single output TA and oscillators using a single input single output TA are summarized.

References

[1] Soliman, A.M. (1979). A New Active C differential input integrator using the DVCCS/DVCVS. *International Journal of Circuit Theory and Applications* vol. 17(2), 272-275.

[2] Geiger R.L.& Sanchez-Sinenciio E. (1985). Active filter design using operational transconductance amplifiers: A tutorial, *IEEE Circuits Devices Magazine* vol. 1(2), 20–32.

[3] Geiger R.L., Sanchez-Sinenciio E. & Lozano H.N. (1988). Generation of continuous time two integrator loop OTA filter structures, *IEEE Trans. on Circuits and Systems* vol. 35(8), 936-946.

[4] Ananda Mohan P.V. (1990). Generation of OTA-C filter structures from active RC filter structures, *IEEE Trans. Circuits and Systems* vol. 37(5), 656–659.

[5] Mahattanakul J.& Toumazou C. (1998).Current mode versus voltage mode Gm-C biquad filters: What the theory says, *IEEE Trans. on Circuits and Systems II* vol. 45(2),173-186.

[6] Tao Y. & Fidler J.K. (2000). Electronically tunable dual-OTA second-order sinusoidal oscillators/filters with non-interacting controls: A systematic synthesis approach, *IEEE Trans. Circuits Systems I* vol. 47(2), 117-129.

[7] Tan M.A. & Schaumann R.(1988). Design of a general biquadratic filter section with only transconductances and grounded capacitors, *IEEE Trans. Circuits Systems* vol. 35(4), 478–780.

[8] Nawrocki R.& Klein U.(1986). New OTA-capacitor realization of a universal biquad, *Electron. Letters* vol. 22(1), 50–51.

[9] Greer N.P.J., Henderson R.K., Ping L. & Sewell J.I.(1994). Matrix methods for the design of transconductor ladder filters, *IEE Proc., Circuits Devices Systems* vol. 141(2), 89–100.

[10] Schaumann R., Ghausi M.S. & Laker K.R.(1990). *Design of Analog Filters Passive, Active RC and Switched Capacitor*, Prentice Hall, New Jersey.

[11] Payne A, Toumazou C. Analog amplifiers: classification and generalization, *IEEE Trans. Circuits and Systems* I, 1996; 43(1), 43-50.

[12] Chang C.M., Al-Hashimi B.M.& Ross J.N.(2004). Unified active filter biquad structures, *IEE Proc. Circuits Devices Systems* vol.151(*4*), 273-277.

[13] Hua W.G.,Fukui Y.,Kubota K.&Watanabe K.(1991). Voltage-mode to current-mode conversion by an extended dual transformation, *IEEE International Symposium on Circuits and Systems, ISCAS* vol. 3, 1833–1836.

[14] Haigh D.G., Tan F.Q.& Papavassiliou C.(2005). Systematic synthesis of active-RC circuit building-blocks, *Analog. Integrated Circuits Signal Processing* vol.43(3), 297–315.

[15] Bruton L.T.(1980). *RC Active Circuits Theory and Design*, Prentice Hall,Inc. New Jersey.

[16] Mitra S.K. (1969). *Analysis and Synthesis of Linear Active Networks*, John Wiley, New York.

[17] Carlin H.J. (1964).Singular network elements, *IEEE Trans. Circuit Theory* vol. 11(1), 67-72.

[18] Awad I.A.& Soliman A.M.(1999). Inverting second-generation current conveyors: the missing building blocks, CMOS realizations and applications, *International Journal of Electronics* vol.86(4), 413-432.

[19] Awad I.A.& Soliman A.M.(2002). On the voltage mirrors and the current mirrors, *Analog Integrated Circuits and Signal Processing* vol. 32(1), 79-81.

[20] Saad R.A.& Soliman A.M.(2008).Use of mirror elements in the active device synthesis by admittance matrix expansion, *IEEE Trans. Circuits Systems I*, vol.55(9), 2726-2735.

[21] Awad I.A.& Soliman A.M.(2000). A new approach to obtain alternative active building blocks realizations based on their ideal representations, *Frequenz* vol. 54(11-12), 290-299.

[22] Soliman A.M.(2010). Synthesis of controlled sources by admittance matrix expansion, *Journal of Circuits Systems and Computers* vol. 19(3), 597-634.

[23] Saad R.A.& Soliman A.M.(2010). A new Approach for Using the Pathological Mirror Elements In the Ideal Representation of Active Devices, *International Journal of Circuit Theory and Applications* vol. 38(2), 148-178.

[24] Soliman A.M.(2009). Applications of voltage and current unity gain cells in nodal admittance matrix expansion, *IEEE Circuits and Systems Magazine* vol. 9(4), 29-42.

[25] Carlosena A. & Moschytz G.S.(1993). Nullators and norators in voltage to current mode transformations, *Int. Journal of Circuit Theory and Applications* vol. 21(4), 421-424.

[26] Director S.W.& Rohrer R. A.(1969). The generalized adjoint network and network sensitivities, *IEEE Trans. Circuit Theory* vol. 16 (3), 318-323.

[27] Bhattacharyya B.B.& Swamy MNS.(1971).Network transposition and its application in synthesis, *IEEE Trans. Circuit Theory* vol.18(5), 394-397.

[28] Soliman A.M. (2009). Adjoint network theorem and floating elements in NAM, *Journal of Circuits Systems and Computers,* vol. 18(3), 597-616.

[29] El-Adawy A.A.& Soliman A.M.(2000). A Low voltage single input class AB transconductor with rail-To-rail input range, *IEEE Transactions on Circuits and Systems I* vol. 47(2), 236-242.

[30] Sanchez-Sinencio E.& Silva-Martivnez J.(2000). CMOS transconductance amplifiers, architectures and active filters: A tutorial, *IEE Proc. Circuits Devices Systems* vol.147(1), 3–12.

[31] Ismail A.M., El-Meteny S.K.& Soliman AM.(1999). A New family of highly linear transconductors based on the current tail differential pair, *Microelectronics Journal* vol. 30 (8),753-767.

[32] Deliyannis T., Sun Y.& Fidler J.K.(1999). *Continuous Time Active Filter Design,* CRC Press, Florida.

[33] Mahmoud S.A.& Soliman AM.(1997). A CMOS programmable balanced output transconductor for analog signal processing, *International Journal of Electronics* vol. 82(6), 605–620.

[34] Kamat D.V. , Ananda Mohan P.V.& Prabhu K.G.(2010).Current-mode operational transconductance amplifier-capacitor biquad filter structures based on Tarmy–Ghausi active-RC filter and second order digital all-pass filters, *IET Proc. Circuits Devices Systems* vol. 4(4), 346-364.

[35] Mahmoud S.A.& Soliman AM.(1999). A New CMOS programmable balanced output transconductor and application to a mixed mode universal filter, *Analog Integrated Circuits and Signal Processing* vol.19 (2), 241-254.

[36] Soliman A.M. (1995). Current conveyors steer universal filter, *IEEE Circuits and Devices Magazine* vol. 11(2), 45-46.

[37] Bhattacharyya B.B., Sundaramurthy M.& Swamy M.N.S.(1981). Systematic generation of canonic sinusoidal RC active oscillators, *IEE Proc. Circuits Devices Systems* vol. 128(3), 114-126.

[38] Park S.W.& Sanchez Sinencio E.(2009). RF Oscillator based on a passive RC bandpass Filter, *IEEE J of Solid State Circuits* vol. 44(11), 3092-3101.

In: Analog Circuits: Applications, Design and Performance
Editor: Esteban Tlelo-Cuautle

ISBN: 978-1-61324-355-8
© 2012 Nova Science Publishers, Inc.

Chapter 5

DESIGN OF CURRENT-FEEDBACK OPERATIONAL AMPLIFIERS AND THEIR APPLICATION TO CHAOS-BASED SECURE COMMUNICATONS

M.A. Duarte-Villaseñor, V.H. Carbajal-Gómez and E. Tlelo-Cuautle
INAOE, Department of Electronics
Luis Enrique Erro No. 1, Tonantzintla,
Puebla. 72840 MEXICO

Abstract

This chapter presents the design automation of different circuit topologies of the current-feedback operational amplifier (CFOA) by applying evolutionary algorithms. Basically, it is shown that the CFOA can be designed by interconnecting two voltage followers (VFs) sandwiched by two current mirrors (CM). Both, the VF and CM are encoded by binary strings to generate all possible circuit topologies until implementing the CFOA with metal-oxide-semiconductor field effect transistors (MOSFETs). To highlight the application of the CFOA, we present the application of the commercially available CFOA AD844, in the design of multiscroll chaos generators and the experimental realization of a chaos-based secure communication system.

Introduction

Enhanced analog signal processing applications are realized thanks to the availability of a big plethora of active devices [1]. Among the most known active devices we can identify the conventional operational amplifier (opamp), the unity-gain cells (UGCs) [2, 3], the current conveyor (CC) with all its generations and multiple outputs [4], the operational transconductance amplifier (OTA) [5], the operational transresistance amplifier (OTRA) [6], and the current-feedback operational amplifier (CFOA) [7]. This last active device is also useful in the implementation of active filters [8], sinusoidal oscillators [9, 10], and chaotic oscillators [11, 12]. On the other hand, the majority of these active devices allow to drive voltage and current signals, so that they are very suitable for mixed-mode circuits realizations

[13, 14]. Furthermore, this chapter shows that the CFOA can be designed by combining voltage followers (VFs [15]) with current mirrors (CMs [16]), or by interconnecting two VFs sandwiched by two CMs [11].

To design all CFOA circuit topologies, we present a design automation methodology based on the interconnection of VFs and CMs [17], which are synthesized by genetic algorithms [18, 19], and can be optimized by evolutionary algorithms [20, 21]. Basically, the starting point is the binary genetic encoding of the VF and CM from nullator and norator descriptions [17, 18, 19, 20], respectively. All kinds of VF and CM circuit topologies can be generated by genetic algorithms as well as their interconnection to synthesize the CFOA [18].

As the CFOA is designed by interconnecting VFs driving voltage signals and CMs driving current signals, it is a mixed-mode active device very suitable for the design of chaotic circuits [11, 14, 22]. Besides, among all kinds of chaos generators, we present the CFOA-based realization of multiscroll chaos generators [23, 24], because they provide more complex dynamical behavior [25], than double-scroll chaos generators [11, 22]. However, in the design of integrated circuits, the active devices present limitations, e.g. operating characteristics of the metal-oxide-semiconductor field-effect-transistors (MOSFETs) [26], and frequency of operation [27], so that we need to search for the optimal circuit topology of the CFOA to enhance the performances of the multiscroll chaos generators, which can be implemented in one or multiple dimensions [28].

It has been demonstrated that using chaos generators, the implementation of secure communication systems [29, 30, 31], can be realized. Besides, the design of a secure communication system requires to synchronize two chaos generators [32]. In this work we present the synchronization of two chaos generators using CFOAs by applying Hamiltonian forms and observer approach [30, 31, 33, 34].

In the next sections we present the binary genetic encoding of the VF and CM, and then the design automation of CFOAs by interconnection of VFs and CMs. Afterwards, the design of multiscroll chaos generators using CFOAs is described. The experimental realization of two multiscroll oscillators using the commercially available CFOA AD844 is also introduced. Finally, the experimental synchronization of two chaotic oscillators by applying Hamiltonian forms and observer approach, is presented in order to implement a chaos-based secure communication system.

Binary Genetic Encoding of Active Devices

This section summarizes how to encode the circuit topology of the VF and CM into binary chromosomes, in order to generate new circuit topologies or solutions from the evolution of populations. For instance, the binary genetic enconding (BGE) of the VF begins from its ideal description using nullators, as already shown in [15]. For synthesis purposes, each nullator is joined with a norator in order to synthesize the pair by a transistor, as shown in Figure 1, where the node with the joined terminals of the nullator (O) and norator (P) elements is associated to the source (S) terminal of a MOSFET, the other terminal of the O element is associated to the gate (G), and the other terminal of the P element to the drain (D). The combinations in forming O-P pairs generate a small-signal gene called genSS. The O-P pairs are biased by addition of voltage and current DC levels, generating a gene called genBias. Every O-P pair can be replaced by an N-channel or P-channel MOSFET generating the gene called genSMOS. Every current source is replaced by a MOSFET-based current mirror, leading to the generation of the gene called genCM.

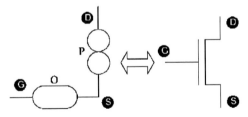

Figure 1. Ideal nullator-norator (O-P) based description of a MOSFET.

As a result, the proposed genetic representation for the VF consists of a chromosome of four ordered genes, as shown by Eq. (1) [15, 18, 19, 20]. By using n as the number of O-P pairs, and m as the number of bits for genCM, the length of Eq. (1) is calculated by Eq. (2). In this manner, by beginning with 1, 2 or 4 O elements [15], if $m=2$, we have 7, 12 or 22 bits to encode a VF.

$$Chromosome_{VF} = genSS * genSMos * genBias * genCM \qquad (1)$$

$$Length\_Chromosome_{VF} = 2n + 1n + 2n + m \qquad (2)$$

The BGE for the CM begins from the norator or P-based descriptions shown in Figure 2 [18]. In this case, it is necessary to add nullators (O elements) to form joined O-P pairs and to generate the gene called genSS. From Figure 1, the MOSFET can be N-channel or P-channel, generating the gene called genSMos. However, in this case, the node *ref* determines the kind of MOSFET, i.e. if ref=0 means that this node is connected to ground or to a negative voltage bias level (e.g. VSS), so that all O-P pairs are synthesized by N-MOSFETs, else, if ref=1 it means that this node is connected to a positive voltage bias level (e.g. VDD), so that all O-P pairs are synthesized by P-MOSFETs. In this case, we do not need to add combinations of current DC bias levels so that the gene called genBias used in Eq. (1) [15], is not present in the genetic encoding of the CM. But, in this case, two current DC bias levels are always connected to nodes *in* and *out* in Figure 2, whose direction is determined according to the connection of node *ref*. On the other hand, the gates of the MOSFETs requires of voltage DC bias levels leading to the generation of the gene called genVbias [18]. This gene consists of 1, 2 and 3 bits for Figure 2(a), 2(b) and 2(c), respectively. In the first case, the gate of a MOSFET can be connected to node *in* or *out* (1 bit). From Figure 2(b), the gate of a MOSFET can be connected to nodes *in, out, y,* and *bias*, where node *bias* is a negative (when ref=0) or positive voltage DC bias level (when ref=1). From Figure 2(c), genVbias needs 3 bits to encode the combinations of the gate of a MOSFET, which can be connected to nodes *in, out, x, y,* and *bias*. Finally, the BGE of the CM consists of a chromosome of three ordered genes, as shown by Eq. (6). By using n as the number of O-P pairs, the length of Eq. (3) is calculated by Eq. (4). By beginning with 2 (Figure 2(a)), 3 (Figure 2(b)), or 4 (Figure 2(c)) P elements [18], we have 6, 9 or 12 bits to encode the CM.

$$Chromosome_{CM} = genSS * genSMos * genVbias \qquad (3)$$

$$Length\_Chromosome_{CM} = 2n + 1 + (n-1) \qquad (4)$$

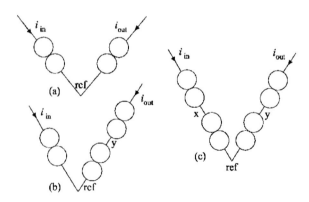

Figure 2. Norator-based descriptions of the CM.

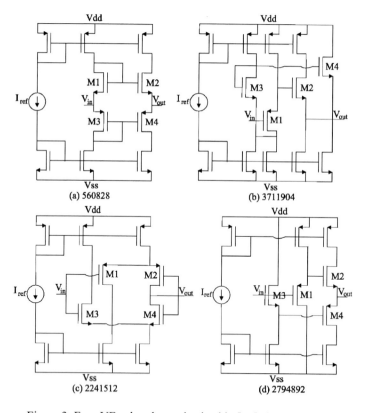

Figure 3. Four VFs already synthesized in [15], from the BGE.

The automatic design automation of VFs and CMs can then be performed by genetic operations [17, 18, 19, 20], and begining with the creation of random solutions according to Eq. (1) for the VF, and Eq. (3) for the CM. For instance, the main steps for the synthesis of VFs are the following [15]:

1. A start-population (Pop_k) consisting of several chromosomes with ordered genes, as shown in Eq. (1), is randomly created.
2. The chromosomes are evaluated by verifying that for genSS and genBias, neither the input-port nor the output-port is connected to the voltage DC bias levels, e.g. either VDD or VSS.
3. If step 2 is accomplished, genSS, genSMos, genBias and genCM are decoded to generate a SPICE netlist to simulate every chromosome.
4. The VF is sized to accomplish the desired objectives.

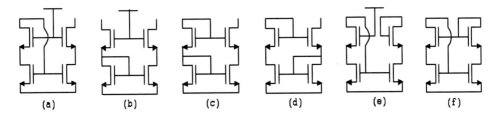

Figure 4. Six CMs already synthesized in [18], from the BGE.

The execution of the genetic algorithm generates different circuit topologies, for example the ones shown in Figure 3 for the VF. In a similar way, Figure 4 shows some CMs synthesized from the chromosomal representation shown in Eq. (3).

A simple behavior of the population beginning with 20 individuals through 16 generations is shown in Figure 5. This is an optimistic case where it can be appreciated that the genetic algorithm maintains an average number of individuals, and it needs relatively a few iterations to select a circuit topology. In Figure 6 is shown the normalized CPU-time, where one can infer that the majority of CPU time is required for the circuit simulation and evaluation of the fitness functions related to the electrical performances to select an optimal circuit topology.

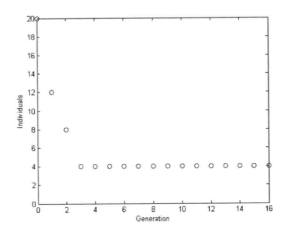

Figure 5. Behavior of the population through 16 generations.

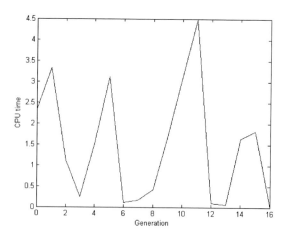

Figure 6. Behavior of the CPU-time through 16 generations.

The synthesized VFs and CMs have exactly unity-gain in the ideal case, infinity bandwidth, infinity/zero input impedance, zero/infinity output impedance, and so on. However, when they are designed with real MOSFETs, their gain is close to unity, the bandwidth is finite and the impedances are not infinity and/or zero. In this manner, the design automation procedure is quite useful to synthesize the optimal circuit topology to accomplish desired target especifications. The key points regarding tracking errors in the design of the VF and CM have been introduced in [26], where it is shown the realization of some dynamical systems using MOSFETs.

Design Automation of CFOAs

In [18, 26] is shown that the interconnection of the VF with CMs leads us to the design of the positive-type second generation current conveyor (CCII+) [1, 4, 13]. In a next step of design, the interconnection of the CCII+ with a VF leads us to the design of the CFOA [1, 4, 7], as shown in Figure 7(a). In this manner, the design automation of the CFOA can be performed by the interconnection of two VFs sandwiched by two CMs, as already demonstrated in [11].

In a more simple way, the design of the CFOA can be realized by the interconnection of three blocks, namely VF-CM-VF, where the VF at the input and output can be the same, as it was done for the commercially available CFOA AD844, which is implemented with bipolar junction transistors (BJTs). However, the connection between the first VF with the CM is not trivial, it requires the superimposing of subcircuits [17], or the modification of the design automation of the VF to create another terminal to implement the CCII+. On the other hand, the connection of the CM with the second VF is done by a simple union or cable. As a result, the BGE of the CFOA can be done by the superimposing of the chromosomes of the first VF with the CM, plus the chromosome of the second VF, as already shown in [18].

Lets us consider the VF shown in Figure 3(d), this topology has been used to design the commercially available CFOA AD844 but using BJTs. If this VF circuit topology is interconnected with simple current mirrors, we obtain the CCII+ shown in the left part of

Figure 7(b). In this case, the VF is embedded between nodes labeled Y and X, and the CM is embedded between nodes labeled X-Z. When another VF is cascade connected, we get the CFOA, as shown in Figure 7(b). In this CFOA circuit topology, the VF between nodes Y-X and between nodes Z-W have the same initial topology shown in Figure 3(d), but the first VF was evolved to create the CCII+ [18].

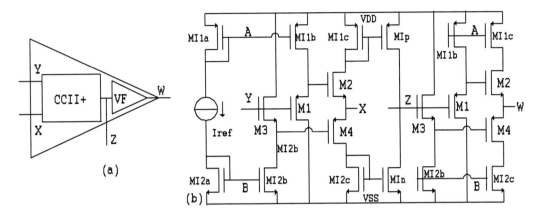

Figure 7. CFOA description by connecting: (a) a CCII+ with a VF, and (b) circuit realization.

As one sees, the CFOA has been designed by the interconnection of the three blocks VF-CM-VF. For instance, by combining the four VFs in Figure 3, with the six CMs in Figure 4, we can design up to 24 CCII+s. Afterwards, by combining the 24 CCII+s with the four VFs in Figure 3, we can design up to 96 CFOA circuit topologies. Besides, from Eq. (1) and (2), if we encode the VF with 22bits, we have a search space of 2^{22}=4,194,304 circuit topologies, and if we encode the CM with 12 bits, we have a search space of 2^{22}=4,096 circuit topologies. The combination of these VF with the CM circuit topologies leads us to a search space of 17,179,869,184 circuit topologies to design the CCII+. Finally, the combination of these CCII+ circuit topologies with the VF circuit topologies leads us to a seach space of 72,057,594,037,927,936 circuit topologies to design the CFOA. Obviously, many of that CFOA circuit topologies are not useful, but to find those ones, we need to apply an intelligent system such as evolutionary algorithms. In Figure 8 and Figure 9 we show other two CFOA circuit topologies designed with gentic algorithms.

Design of Multiscroll Chaos Generators

A chaotic oscillator consists of a linear part and a nonlinear part [23]. In some cases, the nonlinear part can be approached to a piecewise-linear (PWL) function [12, 22, 23, 24, 25, 26, 27, 28], which is relatively easy to implement.

Starting from the description of a continuous-time chaotic oscillator by state variables, its design by using CFOAs realized with MOSFETs, implies the implementation of linear operators such as addition, subtraction, gain, integration and derivation; as well as the nonlinear or PWL functions. In the following sections we show the construction of multiscroll chaos generatos, which are used to design a secure communication system.

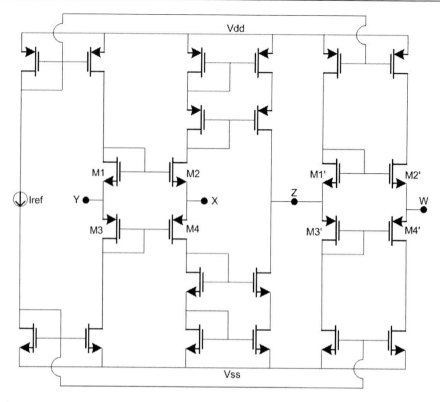

Figure 8. CFOA designed by combining the VF in Figure 3(a) and the CM in Figure 4(c).

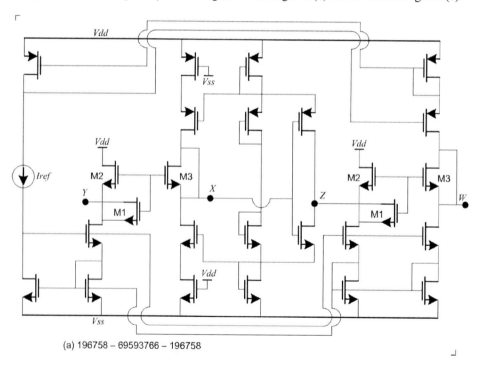

(a) 196758 – 69593766 – 196758

Figure 9. CFOA designed by combining a new VF circuit topology with a new CM topology.

Chaotic oscillators based on saturated nonlinear functions (SNLFs) can be modeled using PWL aproximations. In Figure 10 is shown a SNLF with 5 and 7 segments to generate multiscrolls. In Eq. (5) is described a PWL approximation called series of a SNLF, where $k \geq 2$ is the slope of the SNLF and multiplier factor to saturated plateus, $plateau = \pm nk$ with $n = integer\ odd$ to generate even-scrolls and $n = integer\ even$ to generate odd-scrolls. $h = saturated\ delay$ of the center of the slopes in Figure 10, and must agree with $h_i = \pm mk$, where $i = 1,...,[(scrolls - 2)/2]$ and $m = 2,4,...,(scrolls - 2)$ to generate even-scrolls; and $i = 1,...,[(scrolls - 1)/2]$ and $m = 1,3,...,(scrolls - 2)$ to generate odd-scrolls; p and q are positive integers.

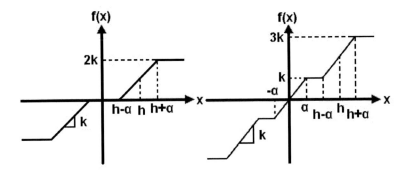

Figure 10. PWL description of a SNLF with 5 and 7 segments.

$$f(x;k,h,p,q) = \sum_{i=-p}^{q} f_i(x;h,k) \qquad (5)$$

To generate multiscrolls attractors (*n*) a non-linear controller is added to the linear system, as shown in Eq. (6), where $f(x;k,h,p,q)$ is defined by Eq. (7), and a,b,c,d are positive constants and must be $0 < a,b,c,d < 1$ to accomplish chaos conditions [35].

$$\begin{aligned} \dot{x} &= y \\ \dot{y} &= z \\ \dot{z} &= -ax - by - cz - df(x;k,h,p,q) \end{aligned} \qquad (6)$$

The simulation of the multiscrolls attractors modeled by Eq. (6) and Eq. (7), is executed using the numerical procedure presented in [36, 37]. 6-scrolls attractors are generated by setting $a = b = c = d = 0.7$, $k = 10, h = 20$, $p = q = 2$, as shown in Figure 11. As one sees, Figure 11(a) shows that the levels of voltage or current signals are very high, this is associated to the dynamic ranges (DRs) that the active devices can provide. In circuit realization, these high levels are downscaled, because real electronic devices cannot handle hign DRs. On the other hand, Eq. (7) cannot be synthesized by electronic circuits and it cannot have small DRs because $k \geq 2$ [35].

$$f(x;k,h,p,q)=\begin{cases} (2q+1)k & if & x>qh+1 \\ k(x-ih)+2ik & if & |x-ih|\leq 1 \\ & & -p\leq i\leq q \\ (2i+1)k & if\ ih+1<x<(i+1)h-1 \\ & & -p\leq i\leq q-1 \\ -(2p+1)k & if & x<-ph-1 \end{cases} \tag{7}$$

Consequently, the following equalities must be accomplished: $h=2k$ or $h=k$ for even or odd scrolls, respectively, to avoid superimposing of the slopes because the plateaus can disappear. Henceforth, the value of α is restricted to 1, so that to implement multiscrolls attractors using practical active devices one needs to downscale the DRs of the SNLFs [35]. Then, for circuit realization purposes, the SNLF series is redefined by Eq. (8), this new relationship allows that $k<1$ because the chaos-condition now applies on $s=k/\alpha$, related to the new slope. In this manner, k and α can be selected to permit that $k<1$, so that the DRs in Eq. (7) can be downscaled. As a result, 6-scrolls attractors can be generated by setting $a=b=c=d=0.7$, $k=1$, $\alpha=6.4e^{-3}$, $s=156.25$, $h=2$, $p=q=2$, as shown in Figure 11(b). Now, the DRs of the attractors are within the DRs allowed by the real active devices. Besides, it is possible to have smaller DRs depending on the values of k and α.

$$f(x;k,h,p,q)=\begin{cases} (2q+1)k & if & x>qh+\alpha \\ \dfrac{k}{\alpha}(x-ih)+2ik & if & |x-ih|\leq\alpha \\ & & -p\leq i\leq q \\ (2i+1)k & if\ ih+\alpha<x<(i+1)h-\alpha \\ & & -p\leq i\leq q-1 \\ -(2p+1)k & if & x<-ph-\alpha \end{cases} \tag{8}$$

Circuit Realization of Multiscroll Chaos Generators

The dynamical system described in Eq. (6) has the block diagram representation shown in Figure 12, which can be realized with 3 integrators and an adder. Each block can be realized with different kinds of active devices, namely: OpAmps [23, 28, 38, 39], CFOAs [11, 12, 40], current conveyors (CCs) [24, 30], unity-gain-cells (UGCs) [18, 26], and so on. For instance, the circuit realization of the dynamical system from Eq. (6) using CFOAs, is shown in Figure 13.

Design of Current-Feedback Operational Amplifiers and Their Application ...

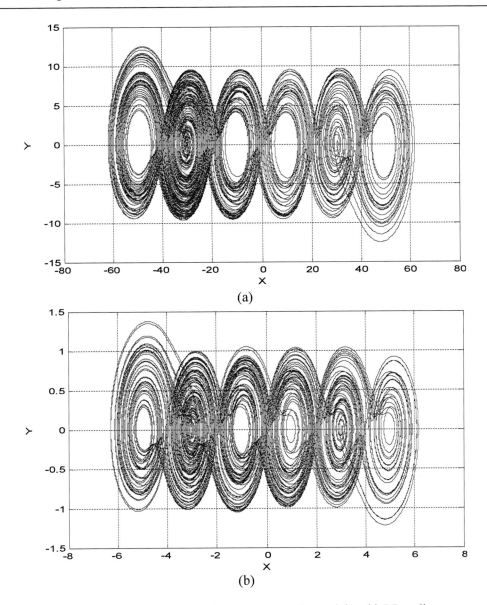

Figure 11. 6-scrolls attractors: (a) without DR scaling and (b) with DR scaling.

By applying Kirchhoff's current-law in Figure 13 one obtains Eq. (9), where $SNLF = i(x)Rix$. The parameters are determined by Eq. (10). The SNLF can be designed by using CFOAs working in the saturation region with shift bias-levels. The CFOA voltage behaviors can be modeled by the opamp finite-gain model shown in Figure 14 [37], so that a SNLF can be described by $V_o = A_v/2 \left(\left| V_i + V_{sat}/A_v \right| - \left| V_i - V_{sat}/A_v \right| \right)$. If a shift-voltage $(\pm E)$ is added, one gets the shifted-voltage SNLFs determined by Eq. (11) for positive and negative shifts, respectively. Now, $\alpha = V_{sat}/A_v$ are the breakpoints, $k = V_{SAT}$ is the saturated plateau, and $s = V_{sat}/\alpha$ is the saturated slope. A resistor can be added to realize a current-to-voltage transformation, e.g. $i_o = V_o/R_C$.

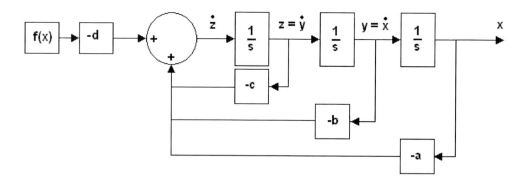

Figure 12. Block diagram description of Eq. (6).

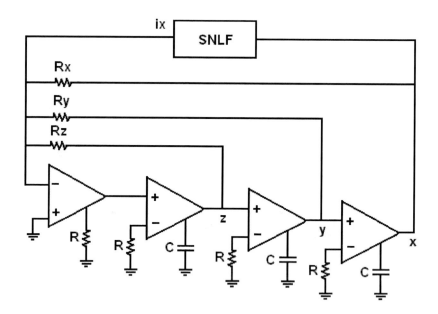

Figure 13. CFOA-based implementation of Eq. (6).

$$\frac{dx}{dt} = \frac{y}{RC}$$
$$\frac{dy}{dt} = \frac{z}{RC} \qquad (9)$$
$$\frac{dz}{dt} = -\frac{x}{RxC} - \frac{y}{RyC} - \frac{z}{RzC} + \frac{i(x)Rix}{RixC}$$

$$C = \frac{1}{0.7 Rix}, \qquad Rx = Ry = Rz = \frac{1}{0.7C}, \qquad R = \frac{1}{C} \qquad (10)$$

To generate the SNLF, E takes different values in Eq. (11) to synthesize the required plateaus and slopes. The cell shown in Figure 15 is used to realize voltage and current SNLFs from Eq. (11). The value of the plateaus k, in voltage and current modes, the breakpoints α, the slope and h are evaluated by Eq. (12) [38].

$$V_o = \frac{A_v}{2}\left(\left|V_i + \frac{V_{sat}}{A_v} - E\right| - \left|V_i - \frac{V_{sat}}{A_v} - E\right|\right) \quad V_o = \frac{A_v}{2}\left(\left|V_i + \frac{V_{sat}}{A_v} + E\right| - \left|V_i - \frac{V_{sat}}{A_v} + E\right|\right) \qquad (11)$$

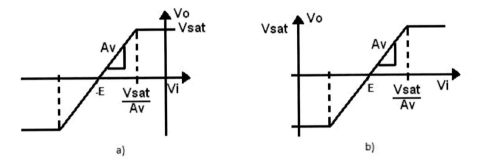

Figure 14. SNLF shifted in voltage: (a) negative shift (b) positive shift.

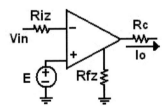

Figure 15. Basic cell to generate SNLFs with the CFOA.

$$k = R_{ix} I_{sat}, \qquad I_{sat} = \frac{V_{sat}}{R_C}, \qquad \alpha = \frac{R_{iz}|V_{sat}|}{R_{fz}}, \qquad s = \frac{h}{\alpha}, \qquad h = \frac{E_i}{\left(1 + \frac{R_{iz}}{R_{fz}}\right)} \qquad (12)$$

The basic cell (BC) shown in Figure 15 can realize the SNLF from Eq. (8), and the number of BCs is determined by BC=(number of scrolls)-1, which are parallel-connected as shown in Figure 16 [38]. Therefore, by setting R_{ix}=10KΩ, C=2.2nf, R=7KΩ, R_x=R_y=R_z=10KΩ, Rf=10KΩ, R_i=10KΩ in Figure 13, and Rix=10KΩ, Rc=64KΩ, Riz=1KΩ, Rfz=1MΩ, E1=±2v with Vsat= +7.24v and -7.28v, the result is N=4-scrolls, F=10Khz and EL=±4v, as shown in Figure 17. The circuit realization was performed by using the commercially available CFOA AD844. The design automation of multiscroll chaos generators for 1-3 dimensions can be found in [23]. Experimental results using CFOAs and current conveyors can be found in [12] and [24], respectively.

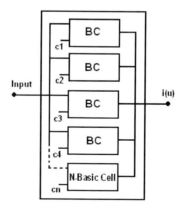

Figure 16. Structure to synthesize the SNLF using the basic cells (BC) realized from Fig. 15.

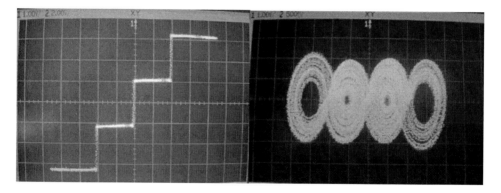

Figure 17. Experimental result of the SNLF to generate 4-scroll attractors.

Synchronization of Two Multiscroll Chaos Generators

In the beginning, a chaotic synchronization method was proposed by Pecora and Caroll [32], and they implicitly propose to create this stable dynamic error system by observing what they called the transverse Lyapunov exponent of the observer. These exponents were used to indicate the stability of the error system. Recently, other implicitly based method which requires no Lyapunov exponent calculation is a Hamiltonian forms and observer approach [33], that we will use to synchronize the multiscroll chaos generator using CFOAs [40].

Lets us consider the dynamic system described by the master circuit in Eq. (13). An slave system is a copy of the master and can be described by Eq. (14).

$$\dot{x} = F(x) \qquad \forall x \in \Re^n \qquad (13)$$

$$\dot{\xi} = F(x) \qquad \forall x \in \Re^n \qquad (14)$$

Design of Current-Feedback Operational Amplifiers and Their Application ...   135

Definition: Two chaotic systems described by a set of states $x_1, x_2, \ldots x_n$ and $\xi_1, \xi_2, \ldots \xi_n$ will synchronize if the following limit fulfills [33, 34, 39]:

$$\lim_{t \to \infty} \left| x(t) - \xi(t) \right| \equiv 0 \tag{15}$$

For any initial conditions $x(0) \neq \xi(0)$.

Due to the real limitations of electronic devices, a tolerance value is used in practical applications, additionally there are some other undesirable agents like noise, distortion, component mismatching, etc [26].

$$\left| x(t) - \xi(t) \right| \leq \varepsilon_t, \qquad \forall t \geq t_f \tag{16}$$

where ε is the allowed tolerance value and a time $t_f < \infty$ is assumed. Eq. (15) and Eq. (16) assume the synchronization error defined as

$$e(t) = x(t) - \xi(t) \tag{17}$$

To satisfy the condition in (15) or (16) between two systems, it is necessary to establish a physical coupling between them through which energy flows. If the energy flows in one direction between the systems, it is one-way coupling, known as master-slave configuration. Furthermore, to synchronize two systems by applying Hamiltonian forms and observer approach, their equations must be placed in the Generalized Hamiltonian Canonical form. Most of the well knew systems can fulfill this requirement, thus, the reconstruction of the state vector from a defined output signal will be possible attending to the observability or detectability of a pair of constant matrices. Some details on Generalized Hamiltonian Systems are summarized herein.

For passive systems, which are a kind of dynamical systems, the energy that flows and dissipates is never more than the energy provided by the source, as a result, a passive system presented in the general form as

$$\dot{x} = f(x) \qquad \forall x \in \mathfrak{R}^n \tag{18}$$

can be fixed in a generalized Hamiltonian form showing a dissipative part and a conservative part as

$$\dot{x} = J(x)\frac{\partial H}{\partial X} + S(x)\frac{\partial H}{\partial X} + F(x) \qquad \forall x \in \mathfrak{R}^n \tag{19}$$

Where $J(x)\dfrac{\partial H}{\partial X}$ is the conservative part and $S(x)\dfrac{\partial H}{\partial X}$ is the non-conservative part which can be dissipative when $S(x)$ is negative or semi-negative definite or considered the destabilizing part for all the other cases (where $S(x)$ is positive, semi-positive definite or undefined sign). $F(x)$ represents the locally destabilizing vector field and $H(x)$ denotes a smooth energy function positive definite in $\Re^n$. The gradient of $H(x)$ is $\dfrac{\partial H}{\partial X}$ and is assumed to exist everywhere. For simplicity, the used energy functions are quadratic forms as shown by

$$H(x)=\frac{1}{2}x^T Mx \tag{20}$$

where M denotes a constant, symmetric, positive definite matrix $\dfrac{\partial H}{\partial X}=Mx$. $J(x)$ and $S(x)$ are also square matrices with the following characteristics:

$$J(x)+J^T(x)=0 \qquad S(x)=S^T(x) \tag{21}$$

Hamiltonian Systems Using Nonlinear Observer

Consider a class of Hamiltonian forms with destabilizing vector field $F(y)$ and lineal output $y(t)$ of the form

$$\dot{x}=J(y)\frac{\partial H}{\partial X}+(I+S)\frac{\partial H}{\partial X}+F(y) \qquad \forall x\in\Re^n; \qquad y=C\frac{\partial H}{\partial x} \qquad \forall y\in\Re^n \tag{22}$$

Where I denotes a constant antisymmetric matrix; S denotes a symmetric matrix; the vector $y(t)$ is the system output and C is a constant matrix. The described system has an observer if one first considers $\xi(t)$ as the vector of the estimated states $x(t)$, when $H(x)$ is the observer's energy function. In addition $N(t)$ is the estimated output calculated from $\xi(t)$ and the gradient vector $\dfrac{\partial H(\xi)}{\partial \xi}$ is equal to $M\xi$ with M being a symmetric constant matrix positive definited. Then, for Eq. (22) a nonlinear observer with gain K is expressed as

$$\dot{\xi}=J(y)\frac{\partial H}{\partial \xi}+(I+S)\frac{\partial H}{\partial \xi}+F(y)+K(y-\eta), \qquad \eta=C\frac{\partial H}{\partial \xi} \tag{23}$$

Design of Current-Feedback Operational Amplifiers and Their Application ...   137

Where the state estimation error is naturally $e(t) = x(t) - \xi(t)$ and the system estimated error output is $e_y = y(t) - \eta(t)$, both described by the dynamical system given by

$$\dot{e} = J(y)\frac{\partial H}{\partial e} + (I + S - KC)\frac{\partial H}{\partial e}, \qquad e \in \Re^n; \qquad e_y = C\frac{\partial H}{\partial e}, \qquad e_y \in \Re^m \quad (24)$$

The following assumption has been made with some abuse of notation $\frac{\partial H(e)}{\partial e} = \frac{\partial H}{\partial x} - \frac{\partial H}{\partial \xi} = M(x - \xi)$. Also, the equivalence $W = I + S$ will be assumed. To maintain stability and to guarantee the synchronization error convergence to zero, two theorems are taken into account.

> **THEOREM 1.** *(Sira-Ramírez & Cruz-Hernández, [33]). The state x(t) of the system in the form of Eq. (22) can be globally, asymptotically and exponentially estimated by the state ξ(t) of an observer in the form Eq.(23), if the pair of matrix (C,W) or (C, S), are observable or at least detectable.*

> **THEOREM 2.** *(Sira-Ramírez & Cruz-Hernández,[33]). The state x(t) of the system in the form of Eq.(22) can be globally, asymptotically and exponentially estimated by the state ξ(t) of an observer in the form Eq.(23), if and only if, a constant matrix K can be found to form the matrix [W-KC]+[W-KC]^T=[S-K]+[S-K]^T=2[S-1/2(KC+C^{T}K^T)] which must be negative definite.*

In the successive, to find an observer for a system in the Hamiltonian form described by Eq.(22), the system will be arranged in the form of Eq.(23), maintaining the conditions in Eq.(20) and Eq.(21), keeping observability or at least detectability and proposing a matrix $y(t)$ such that a gain matrix **K** can be found to achieve the conditions of Theorem 2.

Our proposed schemes for the synchronization of multi-scroll chaos systems of the form in Eq.(23), by using CFOAs is shown in Figure 18. The vector **K** in Eq. (23) is the observer gain and it is adjusted according to the sufficiency conditions for synchronization [33]. By setting Rio = 10kΩ, Rfo = 3.9MΩ and Rko = 22Ω, the SPICE simulation responses are shown in Figure 19. The synchronization error is shown in Figure 20, which can be adjusted with the gain of the observer. The coincidence of the states is represented by a straight line with a unity-slope (identity function) in the phase plane of each state as shown in Figure 21(a). The fast Fourier transform (FFT) of the master generator is shown in Figure 21(b).

The circuit realization of Figure 18 [40], was performed by using the commercially available CFOA AD844. Therefore, by setting Rix=10KΩ, C=2.2nf, R=7KΩ, Rx=Ry=Rz=10KΩ, Rf=10KΩ, Ri=10KΩ, and Rix=10KΩ, Rc=64KΩ, Riz=1KΩ, Rfz=1MΩ, E1=±2v with Vsat=+7.24v and -7.28v in the BC (SNLF), the result is N=4-scrolls, F=10Khz, EL = ±5V as shown in Figure 22. The experimental synchronization result by selecting Rio = 10kΩ, Rfo = 3.9MΩ and Rko=3Ω is shown in Figure 23, the coincidence of the states is represented by a straight line with slope equal to unity in the phase plane for each state.

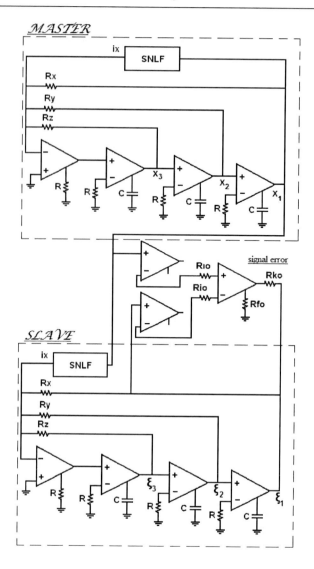

Figure 18. Circuit realization for the synchronization of two multiscroll oscillators in a master-slave configuration using CFOAs.

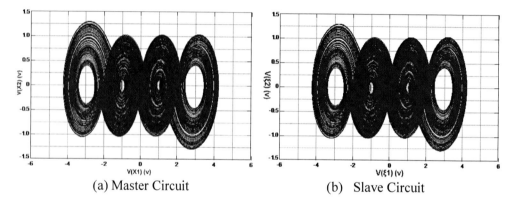

Figure 19. Responses of the master and slave chaos generators with 4-scrolls atractor.

Design of Current-Feedback Operational Amplifiers and Their Application ... 139

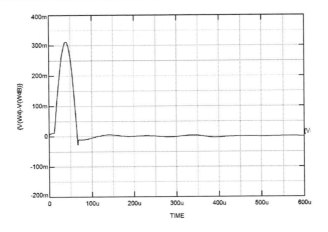

Figure 20. Synchronization Error.

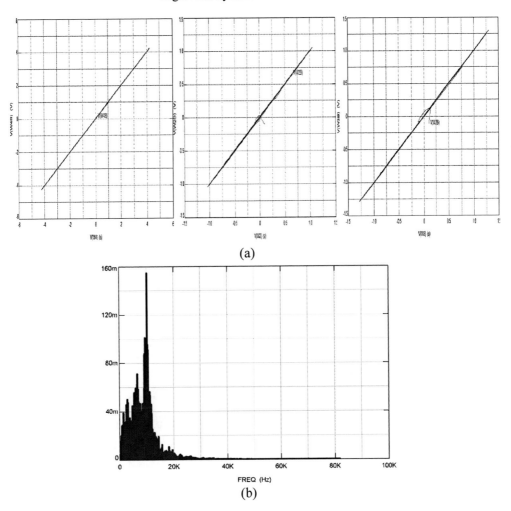

Figure 21. (a) Phase plane synchronization results and (b) FFT.

140   M.A. Duarte-Villaseñor, V.H. Carbajal-Gómez and E. Tlelo-Cuautle

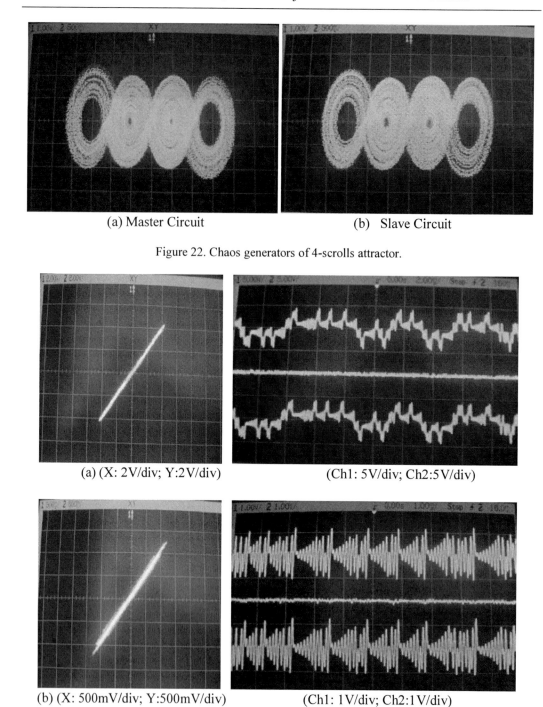

(a) Master Circuit   (b) Slave Circuit

Figure 22. Chaos generators of 4-scrolls attractor.

(a) (X: 2V/div; Y:2V/div)   (Ch1: 5V/div; Ch2:5V/div)

(b) (X: 500mV/div; Y:500mV/div)   (Ch1: 1V/div; Ch2:1V/div)

Figure 25. Continued on next page.

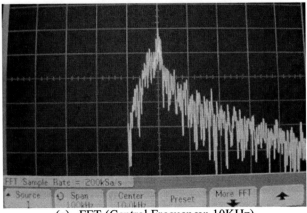

(c) FFT (Central Frecuency: 10KHz)

Figure 23. Experimental synchronization results: Diagram in the phase plane and time signal (a) X_1 vs ξ_1, (b) X_2 vs ξ_2, (c) FFT.

Chaos-Based Secure Communication System

A simple secure communication system in a master-slave topology is presented in this section. A communication system can be realized by using chaotic signals [31, 32, 33, 34, 39, 40]. Chaos masking systems are based on using the chaotic signal, broadband and look like noise to mask the real information signal to be transmitted, which may be analog or digital. One way to realize a chaos masking system is to add the information signal to the chaotic signal generated by an autonomous chaos system, as shown in Figure 24.

As illustrated in Figure 25, this transmission method is to synchronize the systems in a master-slave configuration by a chaotic signal, $x_1(t)$, transmitted exclusively on a single channel, while to transmit a confidential message $m(t)$, it is encrypted with another chaotic signal, $x_2(t)$ by an additive process, this signal can be send through a second transmission channel.

Message recovery is performed by a reverse process, in this case, a subtraction to the signal received $\bar{y} = x_2(t) + m(t)$, it is obvious that we want to subtract a chaotic signal identical to $x_2(t)$ for faithful recovery of the original message. It is important to note that there exists an error in synchrony given by $e_1(t) = x_1(t) - \hat{x}_1(t) = 0$, thus $\hat{m}(t) = m(t)$.

We implemented an additive chaotic masking system using two transmission channels of the form byven by Eq. (22), synchronized by Hamiltonian forms the receiver chaotic system is given by Eq. (23). The communication system is designed using the scenario of unidirectional master-slave coupling, as shown in Figure 26.

The message to convey is a sine wave of frequency f = 10Khz and 500mV in amplitude. Figures 27 and 28 show the experimental result of the secure transmission using chaos generators of 5 and 6-scrolls, respectively. $m(t)$: confidential signal, $x_2(t) + m(t)$: encrypted signal transmitted by the public channel and $\hat{m}(t)$: reconstructed signal by the receiver.

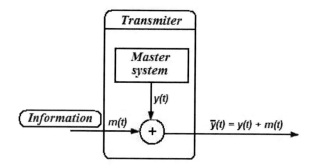

Figure 24. Chaotic masking scheme.

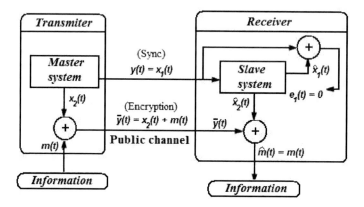

Figure 25. Additive chaotic encryption scheme using two transmission channels.

Conclusion

In the design automation of CFOAs from the described binary genetic encoding (BGE) approach, many circuit topologies can arise. Some of those CFOAs will present good performances after they are optimized. Furthermore, multiscroll chaos generators consisting of PWL functions as nonlinear elements, can be implemented with CFOAs. However, every circuit topology may present drawbacks regarding the operation conditions for the MOSFETs. For instance, we can mention that the active devices implemented with CMOS integrated circuit technology have different frequency ranges of operation, different terminal resistances, different parasitic poles, etc. When the CFOA is used to implement a PWL function, the changing of transistor state can introduce hysteretic effects, because the parasitic capacitance Cgs has slightly changes when the MOSFET goes from saturation to triode, and virtually disappears in the cutoff region.

From the design automation procedure to create different CFOA topologies, we can select the most appropriate to enhance the design of the chaos-based secure communication system introduced heren and realized experimentally with the commercially available CFOA AD844. As shown in this chapter, Hamiltonian forms and observer approach is quite useful two synchronize two chaos generators designed with CFOAs. Finally, experimental results demonstrate the usefulness of the CFOA in implementing communication systems.

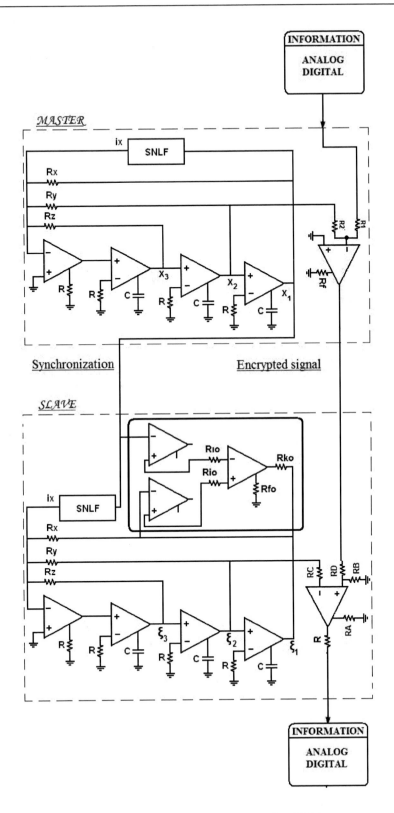

Figure 26. Chaotic transmission system using CFOAs.

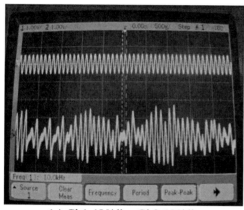
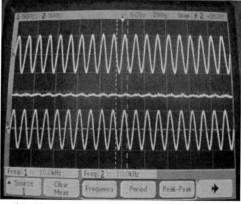

(a) Ch1:1V/div; Ch2:1V/div
Ch1 :$m(t)$, Ch2 : $x_2(t) + m(t)$

(b) Ch1:500mV/div; Ch2:500mV/div;
Center:500mV/div

Figure 27. Chaotic 5-scrolls attractor (a) Encryption of information, (b) Information retrieval.

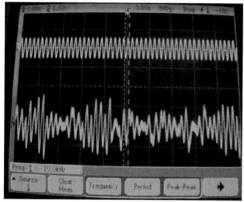
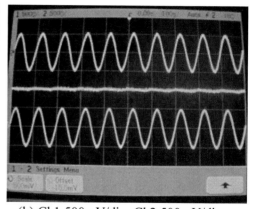

(a) Ch1:1V/div; Ch2:1V/div
Ch1 :$m(t)$, Ch2 : $x_2(t) + m(t)$

(b) Ch1:500mV/div; Ch2:500mV/div;
Center:500mV/div

Figure 28. Chaotic 6-scrolls attractor (a) Encryption of information, (b) Information retrieval.

Acknowledgments

We acknowledge the support from CONACyT under the project number 131839.

References

[1] Sánchez-López, C., Fernández, F.V., Tlelo-Cuautle, E., Tan, S.X.-D. (2011) Pathological element-based active device models and their application to symbolic analysis, *IEEE Transactions on Circuits and Systems I: Regular papers.* DOI: 10.1109/TCSI.2010.2097696

[2] Soliman, A.M. (2009) Applications of Voltage and Current Unity Gain Cells in Nodal Admittance Matrix Expansion, *IEEE Circuits and Systems Magazine*, vol. 9, no. 4, pp. 29-42.

[3] Soliman, A.M. (2010) Transformation of oscillators using Op Amps, unity gain cells and CFOA, *Analog Integrated Circuits and Signal Processing*, vol. 65, no. 1, pp. 105-114.

[4] Tlelo-Cuautle, E., Sánchez-López, C., Moro-Frías, D. (2010) Symbolic analysis of (MO)(I)CCI(II)(III)-based analog circuits, *International Journal of Circuit Theory and Applications*, vol. 38, no. 6, pp. 649-659.

[5] Assaad, R.S., Silva-Martínez, J. (2009) The Recycling Folded Cascode: A General Enhancement of the Folded Cascode Amplifier, *IEEE Journal of Solid State Circuits*, vol. 44, no. 9, pp. 2535-2542.

[6] Sánchez-López, C., Fernández, F.V., Tlelo-Cuautle, E. (2010) Generalized Admittance Matrix Models of OTRAs and COAs, *Microelectronics Journal*, vol. 41, no. 8, pp. 502-505.

[7] Lopez-Martin, A., Ramirez-Angulo, J., Carvajal, R.G., Acosta, L. (2010) Micropower high current-drive class AB CMOS current-feedback operational amplifier, *International Journal of Circuit Theory and Applications*, DOI: 10.1002/cta.674

[8] Gupta, S.S., Bhaskar, D.R., Senani, R., Singh, A.K. (2009) Inverse active filters employing CFOAs, *Electrical Engineering*, vol. 91, no. 1, pp. 23-26.

[9] Gupta, S.S., Sharma, R.K., Bhaskar, D.R., Senani, R. (2010) Sinusoidal oscillators with explicit current output employing current-feedback opamps, *International Journal of Circuit Theory and Applications*, vol. 38, no. 2, pp. 131-147.

[10] Bhaskar, D.R., Senani, R., Singh, A.K. (2010) Linear sinusoidal VCOs: new configurations using current-feedback-opamps, *International Journal of Electronics*, vol. 97, no. 3, pp. 263-272.

[11] Tlelo-Cuautle, E., Gaona-Hernández, A. & García-Delgado, J. (2006). Implementation of a chaotic oscillator by designing Chua's diode with CMOS CFOAs. *Analog Int Circ Signal Proc*, vol 48(2), 159-162.

[12] Trejo-Guerra, R., Sánchez-López, C., Tlelo-Cuautle, E., Cruz-Hernández, C., Muñoz-Pacheco, J.M. (2010) Realization of multiscroll chaotic attractors by using current-feedback operational amplifiers, *Revista Mexicana de Física*, vol. 56, no. 4, pp. 268-274.

[13] Tlelo-Cuautle, E., Sánchez-López, C., Martínez-Romero, E., Tan, S.X.-D., Li, P., Fernández, F.V., Fakhfakh, M. (2011) *Behavioral modeling of mixed-mode integrated circuits,* in Advances in Analog Circuits, Tlelo-Cuautle, E. (Ed.) INTECH.

[14] Klomkarn, K., Sooraksa, P., Chen, G. (2010) New construction of mixed-mode chaotic circuits, *International Journal of Bifurcation and Chaos,* vol. 20, no. 5, pp. 1485-1497.

[15] Tlelo-Cuautle, E., Duarte-Villaseñor, M. A. & Guerra-Gómez, I. (2008). Automatic synthesis of VFs and VMs by applying genetic algorithms, *Circuits, Systems and Signal Processing*, Birkhäuser: Boston, MA, Vol. 27, no. 3, 391-403.

[16] Tlelo-Cuautle, E., Sánchez-López, C., Martínez-Romero, E., Tan, S. X.-D. (2010). Symbolic analysis of analog circuits containing voltage mirrors and current mirrors, *Analog Integrated Circuits and Signal Processing*, vol. 65, no. 1, pp. 89-95.

[17] Tlelo-Cuautle, E., Moro-Frías, D., Sánchez-López, C. & Duarte-Villaseñor, M. A. (2008). Synthesis of CCII-s by superimposing VFs and CFs through genetic operations, *IEICE Electronics Express*, vol 5(11), 411-417.

[18] Tlelo-Cuautle, E. & Duarte-Villaseñor, M. A. (2008). Evolutionary electronics: automatic synthesis of analog circuits by GAs, In *Success in Evolutionary Computation, Series: Studies in Computational Intelligence*; Editor Yang, A., Shan, Y. Bui, & L. T. Springer-Verlag, Vol. 92, 165-188.

[19] Tlelo-Cuautle, E., Guerra-Gómez, I., Reyes-García, C. A. & Duarte-Villaseñor, M. A. (2010). Synthesis of Analog Circuits by Genetic Algorithms and their Optimization by Particle Swarm Optimization, In *Intelligent Systems for Automated Learning and Adaptation: Emerging Trends and Applications*, Editor Chiong, R., IGI Global: Hershey, PA, 173-192.

[20] Tlelo-Cuautle, E., Guerra-Gómez, I., Duarte-Villaseñor, M.A., de la Fraga, L.G., Flores-Becerra, G., Reyes-Salgado, G., Reyes-García, C.A., Rodríguez-Gómez, G. (2010) Applications of evolutionary algorithms in the design automation of analog integrated circuits, *Journal of Applied Sciences*, vol. 10, no. 17, pp. 1859-1872.

[21] Tlelo-Cuautle, E., Guerra-Gómez, I., de la Fraga, L.G., Flores-Becerra, G., Polanco-Martagón, S., Fakhfakh, M., Reyes-García, C.A., Rodríguez-Gómez, G., Reyes-Salgado, G. (2011) *Evolutionary Algorithms in the Optimal Sizing of Analog Circuits*, in Intelligent Computational Optimization in Engineering: Techniques & Applications. M. Koeppen, G. Schaefer, A. Abraham and L. Nolle (Eds.), Springer.

[22] Sánchez-López, C., Muñoz-Pacheco, J.M., Trejo-Guerra, R., Carbajal-Gómez, V.H., Ramírez-Soto, C., Echeverría-Solís, O.S., Tlelo-Cuautle, E. (2011) *Design and applications of continuous-time chaos generators*, in Chaos Systems, Tlelo-Cuautle, E. (Ed.) INTECH.

[23] Muñoz-Pacheco, J.M., Tlelo-Cuautle, E. (2010) *Electronic design automation of multi-scroll chaos generators,* Bentham Sciences Publishers Ltd.

[24] Sánchez-López, C., Trejo-Guerra, R., Muñoz-Pacheco, J.M., Tlelo-Cuautle, E. (2010) N-scroll chaotic attractors from saturated functions employing CCII+s, *Nonlinear Dynamics,* vol. 61, no. 1-2, pp. 331-341.

[25] Trejo-Guerra, R., Tlelo-Cuautle, E., Muñoz-Pacheco, J.M., Sánchez-López, C., Cruz-Hernández, C. (2010) On the Relation between the Number of Scrolls and the Lyapunov Exponents in PWL-functions-based n-Scroll Chaotic Oscillators, *International Journal of Nonlinear Sciences and Numerical Simulation,* vol. 11, no. 11, pp. 903-910.

[26] Trejo-Guerra, R., Tlelo-Cuautle, E., Muñoz-Pacheco, J.M., Cruz-Hernández, C., Sánchez-López, C. (2010) *Operating characteristics of MOSFETs in chaotic oscillators,* in Transistors: Types, Materials and Applications. Benjamin M. Fitzgerald (Ed.), NOVA Science Publishers Inc.

[27] Sánchez-López, C., Muñoz-Pacheco, J.M., Tlelo-Cuautle, E., Carbajal-Gómez, V.H., Trejo-Guerra, R. (2011) On the trade-off between the number of scrolls and the operating frequency of the chaotic attractors, *IEEE International Symposium on Circuits and Systems (ISCAS)*, Rio de Janeiro, Brasil, May 15-18.

[28] Muñoz-Pacheco, J. M. & Tlelo-Cuautle, E. (2009). Automatic synthesis of 2D-n-scrolls chaotic systems by behavioral modeling. *J Appl Res Tech*, vol 7(1), 5-14.

[29] González, O. A., Han, G., Pineda de Gyvez J. & Sánchez, S. E. (2000). Lorenz-Based Chaotic Cryptosystem: A Monolithic Implementation. *IEEE Trans Circ Sys Fund Theor Appl, vol* **47**(8), 1243-1247.

[30] Trejo-Guerra, R., Tlelo-Cuautle, E., Cruz-Hernández, C. & Sánchez-López, C. (2009). Chaotic communication system using Chua's oscillators realized with CCII+s. *Int J Bifurcat Chaos, Appl Sci Eng.*, vol 19(12).

[31] Cruz-Hernández, C., López-Mancilla, D., García-Gradilla, V., Serrano-Guerrrero, H. & Núñez-Pérez, R. (2005). Experimental Realization of Binary Signals Transmission Using Chaos. *J Circ Sys Comput*, vol 14(3), 453-458.

[32] Pecora, L. M. & Carroll, T. L. (1990). Synchronization in chaotic systems. *Phys. Rev. Lett.*, vol 64, 821-824.

[33] Sira-Ramírez, H. & Cruz-Hernández, C. (2001). Synchronization of Chaotic Systems: A Generalized Hamiltonian Systems Approach. *Int J Bifurcat Chaos Appl Sci Eng*, vol 11(5), 1381-1395.

[34] Gámez-Guzmán, L., Cruz-Hernández, C., López-Gutierrez R. M. & García-Guerrero, E. (2009). Synchronization of Chua's Circuits with Multi-Scroll Attractors: Application to Communication. *Comm Nonlinear Sci Numer Simulat*, vol **14**, 2765-2775.

[35] Lü, J., Chen, G., Yu, X. & Leung, H. (2004). Design and analysis of multiscroll chaotic attractors from saturated function series, *IEEE Trans. Circuits Syst. I* **51**(12): 2476–2490.

[36] Tlelo-Cuautle, E. & Muñoz-Pacheco, J.M. (2007). Numerical simulation of chua's circuit oriented to circuit synthesis, *International Journal of Nonlinear Sciences and Numerical Simulation* 8(2): 249–256.

[37] Chen, W., Vandewalle, J. & Vandenberghe, L. (1995). Piecewise-linear circuits and piecewiselinear analysis: *Circuits and filters handbook, CRC Press/IEEE Press.*

[38] Muñoz-Pacheco, J.M. & Tlelo-Cuautle, E. (2008). Synthesis of n-scroll attractors using saturated functions from high-level simulation, *Journal of Physics-Conference Series (2nd Int. Symposium on Nonlinear Dynamics).*

[39] Shuh-Chuan, T., Chuan-Kuei, H., Wan-Tai, C. & Yu-Ren, W. (2005). Synchronization of chua chaotic circuits with application to the bidirectional secure communication systems, *International Journal of Bifurcations and Chaos* 15(2): 605–616.

[40] Carbajal-Gómez, V.H., Tlelo-Cuautle, E., Trejo-Guerra, R. Sánchez-López, C., Muñoz-Pacheco, J.M., Experimental Synchronization of Multiscroll Chaotic Attractors using Current-Feedback Operational Amplifiers, *Nonlinear Science Letters B: Chaos, Fractal and Synchronization* **1**(1): 37-42, 2011.

In: Analog Circuits: Applications, Design and Performance   ISBN 978-1-61324-355-8
Editor: Esteban Tlelo-Cuautle   © 2012 Nova Science Publishers, Inc.

Chapter 6

ANALOG CMOS MORPHOLOGICAL EDGE DETECTOR FOR GRAY-SCALE IMAGES

Luis Abraham Sánchez Gaspariano and Alejandro Díaz Sánchez[*]
Instituto Nacional de Astrofísica, Óptica y Electrónica,
Luis Enrique Erro No. 1, Santa María Tonantzintla.
Puebla, México

Abstract

This chapter deals with the design of analog CMOS maximum and minimum operators which are intended to be used in massive parallel circuit arrays to accomplish morphological gray-scale image processing, which is an analysis tool useful for extracting components of interest within a gray-scale image, e.g. the edges. Most of the morphological edge detection algorithms have been developed in the area of computer vision since computational methods allow to accomplish highly-complex procedures efficientl . However, real-time processing of the image is hard to achieve by computational methods since the image capture and the processing modules are two different parts of the vision-system. On the other hand, hardware approaches are better suited for real-time image processing since they possess dedicated image capture stages in close interaction with subsequent processing stages. Therefore, current-based Winner-Takes-All (WTA) and Loser-Takes-All(LTA) circuits have been implemented in a double poly three metal layers $0.5\mu m$ CMOS technology. The proposed circuits exhibit an area of $0.0086mm^2$ for the WTA and $0.0096mm^2$ for the LTA, a power consumption of $367.96\mu W$, a maximum frequency of operation of 100KHz, and a minimum difference resolved of $250nA$ with 1KHz input signals.

PACS 05.45-a, 52.35.Mw, 96.50.Fm.

Keywords: CMOS Morphological operators, gray-scale image processing, edge detection, morphological filters, WTA/LTA circuits.

AMS Subject Classification 53D, 37C, 65P.

---

[*]E-mail address: luisabraham.sg@gmail.com, adiazsan@inaoep.mx

1. Introduction

Morphological image processing (MIP) is an analysis tool useful for extracting components of interest within an image [1], e.g., the edges. Most of the morphological edge detection algorithms have been developed in the area of computer vision. Computational methods allow to accomplish highly-complex procedures eff ciently such as morphological watersheds, morphological gradient and top-hat transform, to name a few [2]. However, real-time processing of the image is hard to achieve by computational methods since the image capture and the processing modules are viewed as two different parts of the vision-system [3]. On the other hand, hardware approaches are better suited for real-time image processing since they possess dedicated image capture stages in close interaction with subsequent processing stages. This is the case of the cellular neural network (CNN) [4]. In general, a CNN consists of unit-cells called pixels, where pixels are strongly interconnected allowing a massive parallel computation of the input image. Typically, every single pixel is provided with a photocircuit and a processor. In a CMOS implementation of such vision system, the image capture stage at each pixel can be accomplished by a photodiode whose electrical features at its output are delivered to the processing stage of all the neighbor pixels as well as its own [5].

Due to the characteristics inherent to the CNN architecture, the following issues related with the processing circuitry arises [6]: large area requirements, which restrict the size of the input image to be processed; high power consumption because of the large quantity of circuits within the structure; a degraded precision by phenomena such as mismatch, and the need of high processing rate in order to achieve real-time processing. On the one hand, the election of a digital processor presents the advantage of a high-precision and the drawbacks of high-power consumption and large area requirements[7]. On the other hand, precision of an analog circuit is low, but it exhibits a low-power consumption and less area restrictions when compared to digital approaches. Moreover, processing rate can be done faster in an analog approach, since it has no need of a data memory [8]. Because of all those reasons, the design of an analog CMOS morphological processor for edge detection from gray-scale images is the kernel of this chapter.

2. Gray-Scale Erosion and Dilation Filters

MIP is based on the idea that images represent a collection of spatial patterns that can be analyzed by the form they interact with predef ned patterns called structuring elements [9]. For gray-scale images, morphological operators act on functions def ned on a two-dimensional space. Nevertheless, in order to establish a more accessible understanding of these, the basic gray-scale morphological f lters will be presented here for the case of one-dimensional signals.

Before introducing gray-scale morphological operators, it is necessary to review some mathematical concepts and def nitions. Those are presented in the following. The f rst is concerning to the graph of a signal. The graph of a signal can be translated in two directions: horizontally and vertically. According with the mathematical morphology, when both translations are applied together we obtain a morphological translation, which can be

Analog CMOS Morphological Edge Detector for Gray-scale Images

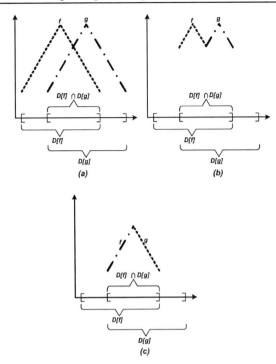

Figure 1. Maximum and minimum operations: (a) f and g, (b) the maximum between f and g, (c) the minimum between f and g.

denoted by:

$$(f_x + y)(z) = f(z - x) + y \tag{1}$$

On the other hand, if g and f are signals with domains D[g] and D[f], respectively, we say that g is beneath f, denoted as $(g \ll f)$, if and only if:

$$D[g] \subset D[f], \text{ and } \forall x \subset D[g], g(x) \leq f(x) \tag{2}$$

Given two arbitrary signals f and g with their respective domains D[f] and D[g], if $x \in D[f] \cap D[g]$, then the minimum and maximum of f and g are denoted as follows:

$$(f \wedge g)(x) = min\{f(x), g(x)\} \tag{3}$$

$$(f \vee g)(x) = max\{f(x), g(x)\} \tag{4}$$

Figure 1 illustrates the concept of minimum and maximum. In (a) we appreciate two arbitrary signals f and g whose respective domains D[f] and D[g], form an intersection. The maximum value between the signals within the intersection is showed in (b) while the minimum in (c).

The ref ection through the origin of a signal h with domain D[h] is denoted by

$$h^{\wedge}(x) = -h(-x) \tag{5}$$

where -x implies ref ection of the signal with respect to the horizontal axis and -h ref ection through the vertical axis.

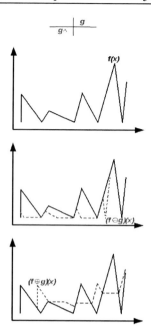

Figure 2. Gray-scale erosion and dilation employing a f at structuring element.

Once we have described the former concepts which are important since they are employed in the conception of morphological operators, we proceed to establish the gray-scale morphological f lters of erosion and dilation. These are the basic operators in mathematical morphology and their iterative use def ne important secondary f lters named opening, closing and hybrid f lters.

The erosion of a signal f by the structuring element g (also a signal) is denoted by (f ⊖ g) and can be formulated by means of the global Minkowski-subtraction as [1]:

$$(f \ominus g) = \wedge \{f_{-x} - g(x) : x \in D[g]\} \qquad (6)$$

In other words, for each point x in the domain of the structuring element g, the morphological translation of $-g$ beneath f is realized, then we take the minimum of this translation.

On the other hand, the dilation of signal f by structuring element g is denoted by (f ⊕ g) and in terms of the of the global Minkowski-addition can be formulated as [1]:

$$(f \oplus g) = \vee \{f_x + g(x) : x \in D[g]\} \qquad (7)$$

Here, for each point $x \in D[g]$, the morphological translation of g over f is attained, then we take the maximum of this translation.

Figure 2, shows the operations of dilating and eroding a signal f by a f at structuring element g. As can be seen, eroding the signal is equivalent to shrinking it while dilating it is equivalent to expanding it.

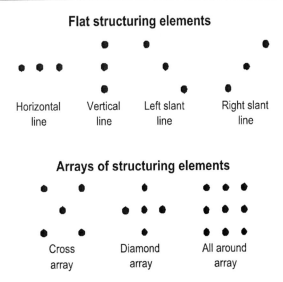

Figure 3. Possible shapes of the structuring elements with the CNN architecture proposed.

3. A Procedure for Detecting Edges Based on Erosion and Dilation

According to the formulations previously presented, the morphological operation of erosion and dilation involve three basic procedures: (1) the election of a structuring element, (2) the morphological translation of the structuring element along the image and (3) the computing of a nonlinear operation. Moreover, those three processes work interactively with each other. Therefore, the structure for the f ltering operation has a profound effect on the performance of the algorithm as a whole.

In the common case that the pixels within the CNN structure are interconnected with all the adjacent neighbors, a 3×3 neighborhood is formed. Assuming that there is control on the activation/deactivation of the pixels from the neighborhood, it is possible to explore the image with seven different structuring elements (SE). Figure 3 shows those possible shapes of SE. As can be seen, both planar and arrays of SE are available. Note that there is symmetry in all possible shapes of SE respect to the central pixel. Consequently, their ref ection through the origin results in the same SE. Thus, morphological translation becomes similar for both erosion and dilation. Therefore, it is possible to attain these two basic morphological operations through the same CNN structure without any extra procedures. In fact, morphological translation is realized in one-step since the array of processors perform simultaneously. Yet again, connectivity is the key for the high processing-rate thus obtained with the CNN structure. Figure 4 illustrates the f ltering of an arbitrary region from the gray-scale image within the CNN structure in case that MIP is realized with the cross array SE.

In order to apply the erosion f lter to the input image, the nonlinear operation of minimum must be compute while the nonlinear operation of maximum has to be applied to the input image in case that the dilation f lter is desired. By applying both erosion and dilation to the input image and then subtracting the dilated and eroded images, the edges of the im-

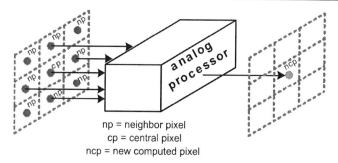

Figure 4. New gray-level computed by the morphological filtering algorithm implemented with the CNN architecture.

age can be extracted. This procedure for detecting edges is referred to as the morphological gradient. It is defined by:

$$GRAD(f) = (f \oplus g) - (f \ominus g) \qquad (8)$$

Figure 5 shows the architecture proposed to accomplish the *morphological gradient algorithm* through the CNN structure. As can be seen, two processors are set at every single pixel. Those correspond to the minimum (erosion) and maximum (dilation) operators. It is important to remark that by adding two processors the structure of the CNN has been modified increasing the area requirements of the processing stage and consequently its power consumption. Therefore, the nonlinear analog processors must be carefully designed to exhibit high-performance within the system. In general terms, to achieve high-performance, the analog circuits employed must be compact, fast, precise and exhibit low-power consumption [6].

4. Implementing Analog CMOS Erosion and Dilation Filters

Once we have established the features of the CNN architecture for boundaries extraction from gray-scale images, we focus our attention on the design of the analog CMOS nonlinear maximum (MAX) and minimum (MIN) circuits. MAX/MIN processors are usually known as Winner-Takes-All and Loser-Takes-All (WTA/LTA) circuits and are inherent operators of nonlinear systems such as nonlinear filters, fuzzy systems and artificial neural systems. Therefore, MAX/MIN operators are widely used in a large variety of nonlinear signal processing tasks, including pattern recognition, classification algorithms and data compression, to name a few. CMOS MAX/MIN circuit realizations are available in both, digital and analog fashion. The analog approaches present some advantages over the digital since analog cells are faster, their area requirements smaller and their power-consumption lower [19]. On the other hand, the precision and resolution of digital cells are higher than those from the analog counterparts.

The operation of a WTA circuit consists of identifying the largest input value among its N external inputs. On the other hand, the operation of a LTA circuit consists of finding the smallest input value among its N external inputs. Depending on their complexity,

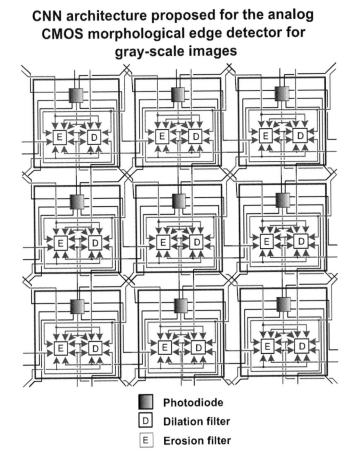

Figure 5. The CNN architecture proposed for implementing the morphological gradient algorithm.

two groups of analog WTA/LTA circuits can be recognized: circuits of quadratic complexity, $O(N^2)$, and circuits of linear complexity, $O(N)$. An $O(N^2)$ complexity circuit is that whose size increases in a quadratic exponential factor according with the number of its N inputs meanwhile an $O(N)$ complexity circuit is that whose size increases linearly according with the number of its N inputs. $O(N^2)$ complexity circuits are typical of those structures capable of comparing only an even number of inputs employing sorting networks. Figure 6 depicts both, quadratic and linear complexity.

On the other hand, according with the manner in which the signals are processed, analog WTAs (LTAs) can be also classifed in three groups: charged-based, current-based and voltage-based processors. Current-based approaches are typically used for processing information directly from sensors. This is the case of many vision chips architectures, where photodetectors typically deliver current-mode signals. Several WTA/LTA architectures have been reported in the literature [10-23]. Of these, the approach proposed by Lazzaro is one of the most employed current-based circuits [10]. The main drawback with this cell is its highly mismatch sensitivity and its low-speed operation. Some authors have tried to improve the circuit of Lazzaro. This is the case of the works reported by DeWeerth [11]

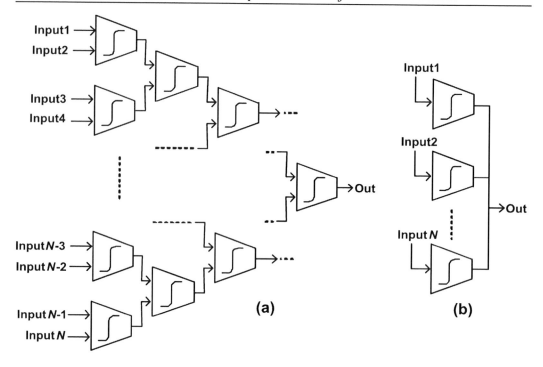

Figure 6. Complexity of WTA/LTA circuits. (a) quadratic complexity and, (b) linear complexity.

and Fish [16], who overcome the disadvantage of low-speed operation. However, despite the improvements obtained on the processing rate, the cell still presents highly mismatch sensitivity.

Figure 7 illustrates a system of linear complexity that realizes both, the MAX and MIN operations. As can be appreciated, it is a model of N cells, such that each cell j, produces an output

$$E_{0j} = \alpha_j U(E_j - E_0) \tag{9}$$

where $j = 1, \ldots, N$

E_j is the external input to the j_{th} cell, $U(\cdot)$ is the step function which can be defned as:

$$U(t - c) = \begin{cases} 0, & t - c < 0 \\ 1, & t - c \geq 0 \end{cases} \tag{10}$$

and

$$E_0 = \sum_{j=1}^{N} E_{0j} = \sum_{j=1}^{N} \alpha_j U(E_j - E_0) \tag{11}$$

A possible solution for (11) can be provided by its intersection with the identity function, $f(E_0) = E_0$, assuming that $\alpha_j > 0 \ \forall j$. In that case, the system has a single solution

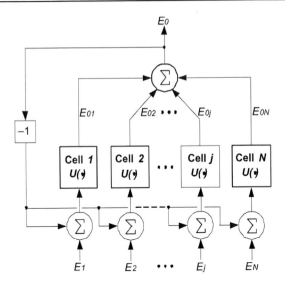

Figure 7. Systematic Model for the Max/Min Operation.

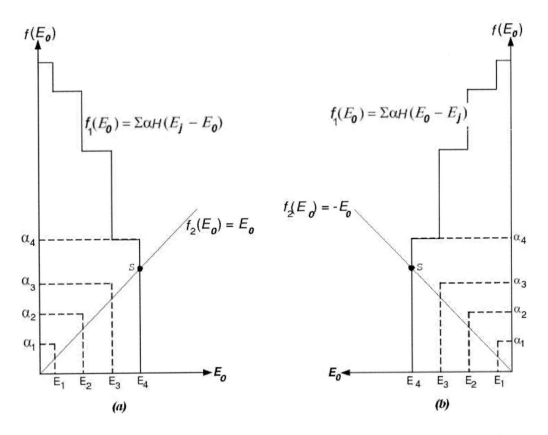

Figure 8. Graphic representation of: (a) the MAX operation and (b) the MIN operation.

and the cell who drives a nonzero output becomes the absolute winner. Therefore, the system behaves just like a (WTA) circuit, and the output corresponds to the maximum value

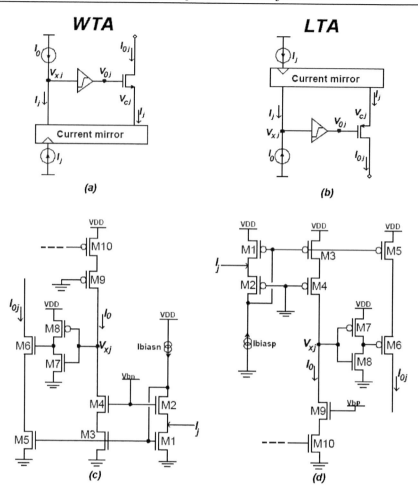

Figure 9. (a) WTA schematic, (b) LTA schematic, (c) WTA circuit diagram, (d) LTA circuit diagram.

from the set of external inputs. An LTA circuit can be obtained if the solution for (11) is given by the negative identity function, $f(E_0) = -E_0$. The graphic representation of the MAX and MIN operations are illustrated on Figure 8.

Figures 9(a) and 9(b) illustrate the schematic diagram of current-based WTA and LTA circuits, respectively. Those implement the operation of the cells previously introduced. They consist of a two-outputs current mirror, a current comparator and a MOS transistor. Each cell j receives two input currents, I_j and I_0, and delivers one output current I_{0j}. In the case of the WTA circuit, if $I_0 > I_j$, the output of the current comparator, v_{0j}, is low and consequently the MOS transistor is turned off. When $I_0 < I_j$, v_{0j} goes high and the MOS is turned on, resulting in $I_{0j} = I_j$. On the other hand, in the case of the LTA circuit, when v_{0j} is low the MOS transistor is turned on and it turns off when v_{0j} is high. Thus, when connecting an N array of WTA cells, the maximum value from a set of N input signals will be compute while the minimum value is obtained with an N array of LTA cells. If the number of cells is too large, high precision in I_0 replication must be guaranteed. In the circuit

Analog CMOS Morphological Edge Detector for Gray-scale Images

Table 1. Dimensions of the transistors from the WTA cell

Transistor	Width (W) $[\mu m]$	Length (L) $[\mu m]$
$M_{1,3,O1,O3}$	30.3	1.2
$M_{2,O2}$	15.6	1.2
$M_{4,O4}$	15.9	1.2
M_5	29.1	1.2
M_6	27.6	1.2
M_7	6.0	0.6
M_8	18.0	0.6
M_9	57.6	1.2
$M_{10,O8,O9}$	45.3	1.2
M_{O5}	56.4	1.2
M_{O7}	58.5	1.2

reported by T. Serrano [22], several simple current mirrors accomplished the replication of I_0 obtaining a good precision without relying on the matching of a large array of transistors. However, the mismatch introduced by the use of simple current mirrors may become unacceptable. Therefore, an improvement on the circuits proposed is the employment of current mirrors based on Flipped Voltage Followers for all the current structures within the circuits since those exhibit a better systematic current matching [24] and thus an improvement on the precision of the current replication can be achieved. Figures 9(c) depicts the circuit diagram of the WTA proposed meanwhile Figure 9 (d) shows the circuit diagram of the LTA. On the other hand, Figure 10 (a) shows the schematic diagram of an N inputs WTA circuit while Figure 10 (b) the schematic of an N inputs LTA circuit. In both cases an output stage must be employed in order to provide a common node for the interconnection of all the cells. Moreover, the output stage establishes a feedback loop by which inhibitory behavior of the corresponding cells in the system is settled.

Both, the WTA and LTA processors have been design in a double poly three metal layers $0.5\mu m$ CMOS technology from MOSIS foundry. For a bias current of $50\mu A$ and a voltage supply of 1.8V, transistor dimensions reported in tables 1 and 2 were employed for WTA and LTA cells, respectively.

In order to evaluate the performance of the circuits if fuctuations on both, process and electrical parameters occur, H-SPICE simulations with Monte-Carlo analysis have been done. Figure 11(a) and (b) show the results obtained for a 5 inputs WTA and a fve inputs LTA circuits, respectively. Thirty iterations were performed varying the threshold voltage V_{TH} and the width of the transistor's channel W. The mismatch model and the Monte-Carlo analysis employed correspond to those proposed in [25]. The black solid lines correspond to the outputs of the circuits while the dotted lines to the inputs. It can be appreciated that the circuits still behave appropriately even with the introduced variations. Therefore, the current-based WTA and LTA exhibit an acceptable robustness.

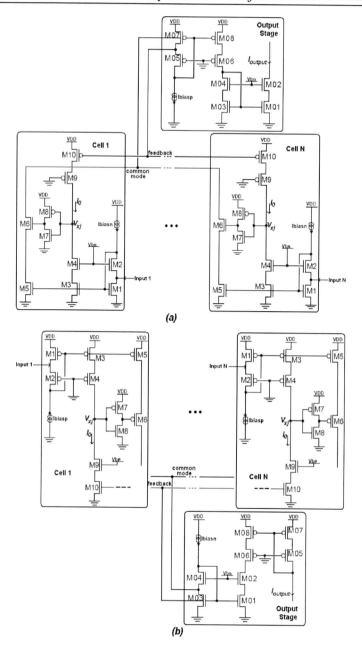

Figure 10. (a) N inputs WTA and (b) N inputs LTA circuits.

5. Results

In this section, experimental results obtained from the WTA and LTA circuits designed under the considerations described previously are presented. In addition, the simulation results obtained from applying the morphological operators of erosion and dilation as well as the morphological gradient algorithm to a gray-scale image by means of those processors are reported.

Table 2. Dimensions of transistors from the LTA cell

Transistor	Width (W) $[\mu m]$	Length (L) $[\mu m]$
$M_{1,3,5,O5,O6}$	45.3	1.2
$M_{2,O4,O7}$	56.4	1.2
M_4	57.6	1.2
M_6	55.2	1.2
M_7	18.0	0.6
M_8	6.0	0.6
M_9	15.9	1.2
$M_{10,O1}$	30.3	1.2
M_{O2}	15.6	1.2
M_{O3}	29.4	1.2

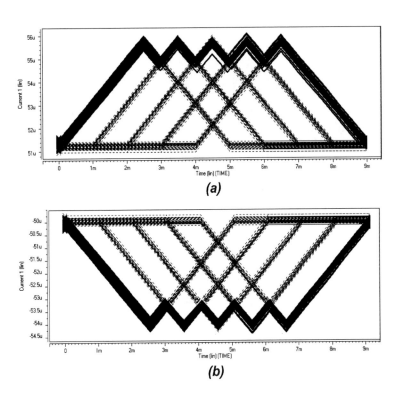

Figure 11. (a) Monte-Carlo Analysis of a fve inputs WTA and (b) Monte-Carlo Analysis of a fve inputs LTA.

5.1. The CMOS WTA and LTA Circuits

Figure 12 shows the photograph of the chip where the WTA and LTA circuits were fabricated. The realization of the system was carried out in a double poly three metal layers $0.5\mu m$ CMOS technology from MOSIS. Each processor has two basic cells and their correspondent output stages. Voltage-to-current converters at the inputs were also included

Figure 12. (a) Chip photograph of the WTA and LTA circuits implemented in a $0.5\mu m$ CMOS technology.

since the waveform generators employed can deliver only voltage signals.

Evaluations on the functionality of both, the WTA and the LTA circuits have been done with a triangular- shape and a sinusoidal-shape signals. Figure 13 (a) shows the results obtained for the case of the WTA meanwhile Figure 13 (b) the results of the LTA. As can be appreciated, the circuits are capable of computing the maximum and minimum, respectively, from the two input signals. It can be noticed certain error when computing the corner-shapes of the signals. This is due to the closeness of the values of the signal in the corner areas and it is a consequence of the f nite resolution of the circuit.

Table 3 presents a summary of the features of the proposed WTA and LTA circuits obtained from the corresponding characterization realized. According to these results, the WTA exhibits an area of $0.0086mm^2$ and the LTA an area of $0.0096mm^2$. In addition, a power consumption of $367.96\mu W$, a maximum frequency of operation of 100KHz and a minimum difference solved of $250nA$ with 1KHz input signals was achieved by the circuits. By analyzing these results with respect to the f gures of merit required for morphological f l-tering (compactness, low power consumption, high processing rate and good precision), we appreciate that the circuits proposed and implemented exhibit a satisfactory performance.

Table 4 presents the characteristics of some reported current-mode WTA/LTA approaches in CMOS technology [10-17]. When compared with other approaches, the one presented here exhibits a higher precision and a good compactness as well as a smaller power consumption and a fast processing rate. It is important to remark the fact that those architectures were implemented in different technologies, and thus, some f gures of merit such as the size of the cell, the power consumption and the processing rate are related to the inherent characteristics of the technology in which the circuits were fabricated. However, the precision achieved by the WTA/LTA proposed is a consequence of the good matching (see Figure 11) of the circuits used to implement the nonlinear functions of MAX and MIN.

Analog CMOS Morphological Edge Detector for Gray-scale Images   163

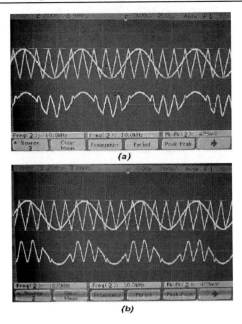

Figure 13. Experimental results obtained for: (a) the maximum operation performed with the WTA and (b) the minimum operation performed with the LTA.

Table 3. Important features of the designed circuits

Figure of merit	WTA cell	LTA cell
# of transistors per cell	10	10
Size of the cell [mm^2]	0.0086	0.0098
Supply [V]	1.8	1.8
Power-consumption [μW]	367.96	367.96
Max. frequency [KHZ]	100	100
Min. diff. resolved @ $1KHZ$	250nA	250nA
Min. diff. resolved @ $100KHZ$	2.3μA	2.3μA
Input swing range	250nA-50μA	250nA-50μA
Hysteresis voltage[mV]	200	8.5
Processing rate [$nsec$]	9.3	12

5.2. Performance of the Analog CMOS Morphological Operators at Image Level

In order to asses the performance of the MAX/MIN operators previously described at image level. The morphological operations of erosion, dilation and the gradient algorithm were carried out with the WTA/LTA circuits proposed within a CNN array like the one depicted in Figure5. The features of the 256 gray-levels image employed are presented in Figure 14. In (a) we can see the original image whose size is 264×365 pixels. On the other hand, the histogram in (b) illustrates the distribution of the intensities along the image. It

Table 4. Some characteristics from different WTA/LTA circuits in CMOS technology

Author	Technoloy [μm]	Min. Diff. solved [μA]	Size [mm^2]	Power [mW]	Supply [V]	Proc. rate [nsec]
Lazaro (1989)	2	2	0.013	5.09	5.0	30-50
DeWeerth (1995)	N.R.	2	N.R.	N.R.	5.0	5-15
Serrano (1995)	2	1	0.070	1.96	5.0	30-500
Demosthenous (1996)	2.4	< 1	N.R.	0.15	5.0	110
Siskos (1999)	1.2	5	N.R.	N.R.	3.0	30
Gwo-Jeng (2000)	0.6	16.5	N.R.	0.125	5.0	25
Poikonen (2002)	0.18	1	N.R.	N.R.	N.R	100-300
Chien Cheng (2003)	0.35	< 1	N.R.	N.R.	3.3	N.R.
Fish (2005)	0.35	5	0.0006	0.09	3.3	12
Ramírez (2005)	0.5	N.R.	0.0817	N.R.	1.5	N.R.
This work	0.5	0.25	0.0086	0.368	1.8	9.3

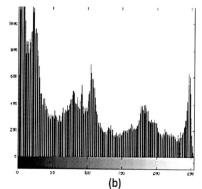

(a) (b)

Figure 14. Features of the gray-scale image employed: (a) original image (264 × 365 pixels), (b) distribution of intensities over the image.

can be noticed that it covers the entire range from 0 to 255 value. Furthermore, the image contains a lot of f ne details as well as a large diversity of contrasts. The face of the girl exhibits rounded shapes upon the f rst plane and some characters can be recognized at the background. Thus, the image has interesting characteristics to evaluate the performance of the proposed nonlinear operators.

The image was morphologically computed as follows: the original image was read with MATLAB and then it was exported to HPSICE where it was processed by the WTAs/LTAs in the way prescribed by the all around structuring element (see Figure 3). Then, the processed image was returned to MATLAB in order to be displayed.

As mentioned earlier, the erosion and dilation f lters as well as the gradient algorithm were applied to the image. Figure 15 depicts the results obtained. As can be appreciated, in the case of the eroded image the contrasts have been darkened and in the case of the

Analog CMOS Morphological Edge Detector for Gray-scale Images 165

Figure 15. (a) eroded image, (b) dilated image, and (c) edge detection.

dilated image the illumination was increased. On the other hand, the edges extracted by means of the gradient algorithm are satisfactory in the sense that the shape of both, coarse and f ne details within the processed image can be differentiated. Details from the face and silhouette of the girl can be distinguished as well as f ne details from the background plane with fairly good clarity.

On the one hand, the main advantage of realizing the edge detection by means of the gradient procedure lies on the simplicity of the algorithm. There are many other approaches to extract the edges of an image with a higher precision. However, most of those methods are only possible to carry out with a computational approach and thus, real-time processing becomes a bottleneck. On the other hand, the main disadvantage of the spatial method used to process the image with the circuits proposed is that it is impractical if the size of the image is too big. However, it is a good option in systems where real-time edge detection of a small image must be done.

6. Conclusion

The design and implementation of current-mode WTA/LTAs circuit of linear complexity in a double poly three metal layers $0.5\mu m$ CMOS technology have been presented. The proposed circuits exhibit an area of $0.0086mm^2$ for the WTA and $0.0096mm^2$ for the LTA, a power consumption of $367.96\mu W$, a maximum frequency of operation of 100KHz, and a minimum difference resolved of $250nA$ with 1KHz input signals. These features provide to the cells proposed with a high precision, a good compactness, a small power consumption and a fast processing rate. From those characteristics, the precision achieved is a consequence of the good matching of the circuits used to implement the nonlinear functions of MAX and MIN. In addition, the basic morphological operations of erosion and dilation as well as the gradient algorithm have been performed with the WTA/LTAs processors. A 264×365 pixels and 256 gray-levels image was eroded, dilated and its edges were extracted. The results attained indicate that modif cations on the illumination and sharpness of the image arise when the erosion and dilation f lters are applied. On the other hand, when the edge detection is carried out by means of the gradient algorithm with an all around structuring element, the shape of coarse and f ne details within the processed image can be distinguished.

A signif cant contribution is attained in this chapter towards the development of continuous-time morphological vision chips since secondary morphological f lters such as the gradient algorithm have been proposed for edge detection with an analog architecture.

References

[1] E. R. Dougherty and J.Astola. *An Introduction to Morphological Image Processing.* SPIE, Washington, 1994.

[2] E. R. Dougherty. *Hands-on Morphological Image Processing.* SPIE, Washington, 2003.

[3] C. J. Lambrecht, editor. *Vision Models and Applications to Image and Video Processing.* Kluwer, Boston, 2001.

[4] A. Moini. *Vision Chips.* Kluwer, Boston, 2000.

[5] T. Roska and A. Rodríguez. *Towards the Visual Microprocessor, VLSI Design and the use of Cellular Neural Networks (CNN) Universal Machine Computers.* John Wiley & Sons, New York, 2001.

[6] M. C. Ornellas and R. Boomgaard. Developing Morphological Building Blocks: From Design to Implementation. In *Proc. IEEE (SIBGRAP 1999),* 1999.

[7] R. C. Gonzalez and R. E. Woods. *Digital Image Processing*. Prentice Hall, New Jersey, 2002.

[8] T. G. Morris and S. P. Deweerth. Analog VLSI Morphological Image Processing Circuit. *IEEE Electronics Lett.*, vol. 31, Nov 1995, No. 23.

[9] P. Soille. Morphological Image Analysis, Principles and Applications. Springer, Berlin, 1999.

[10] J. Lazzaro, S. Lyckeybusch, M. A. Mahowald and C. Mead. Winner-Take-All networks of $O(N)$ Complexity. In *Advances in Neural Information Processing Systems*, 1989.

[11] S. P. DeWeerth and T. G. Morris. CMOS Current Mode Winner-Take-All Circuit with Distributed Hysteresis. *IEEE Electronics Lett.*, vol. 31, Jun 1995, No.13.

[12] T. Serrano and B. Linares. A Modular Current-Mode High-Precision Winner-Take-All Circuit. *IEEE Trans. Circuits Syst. II,* vol. 42, Feb 1995, No. 2.

[13] A. Demosthenous, R. Akbari, S. Smedley and J. Taylor. Enhanced Modular CMOS Current-Mode Winner-Take-All Network. In *Proc. IEEE International Conference on Electronics Circuits and Systems (ICECS 1996)*, 1996.

[14] S. Vlassis, K. Doris and S. Siskos. *Analog Implementation of an Order Statistics Filters. IEEE Trans. Circuits Syst. I*, vol. 46, Oct. 1999, No. 10.

[15] J. Poikonen and A. Paasio. Implementing Grayscale Morphological Operators with a Compact Ranked Order Extractor Circuit. In proc. *IEEE* (*CNNA 2002*), Jul. 2002.

[16] A. Fish, V. Milrud and O. Yadid. High-Speed and High-Precision Current Winner Take All Circuit. *IEEE Trans. Circuits Syst. II*, Jan 2005.

[17] J. Ramírez, G. Doucoudray, R. Carvajal and A. López. Low Voltage High Performance Voltage Mode and Current Mode WTA Circuits Based on Flipped Voltage Followers. *IEEE Trans. Circuits Syst. II: Express briefs*, Vol. 52, July 05.

[18] J. Silva and M. Barranco and E. Sánchez. Modular CMOS charge based Hamming Neural Network. In *Proc. IEEE First International Conference on Electronics Circuits and Systems (ICECS 1994)*, 1994.

[19] Z. Sergin and E. Sánchez. CMOS Winner-Take-All Circuits: A Detail Comparison. In *Proc. IEEE International Symposium on Circuits and Systems (ISCAS 1997)*, 1997.

[20] S. Siskos, S. Vlassis and I. Pitas. Analog Implementation of Fast MIN/MAX Filtering. *IEEE Trans. Circuits Syst. II*, vol. 45, Jul 1998, No. 7.

[21] I. E. Opris. Rail to Rail Multiple-Input Min/Max Circuit. *IEEE Trans. Circuits Syst. II*, vol. 45, Jan 1998, No. 1, 1998.

[22] T. Serrano and B. Linares. A High-Precision Current-Mode WTA-MAX Circuit with Multichip Capability. *IEEE J. Solid-State Circuits*, vol. 33, Feb 1998, No. 2.

[23] J. Ramírez, R.G. Carvajal, A. Torralba, J. Galan, A.P. Vega-Leal and J. Tombs. The Flipped Voltage Follower: A useful cell for low-voltage low-power circuit design. In *Proc. IEEE International symposium on Ciircuit and Systems (ISCAS 2002)*, 2002.

[24] Ramírez-Ángulo, J., Carvajal, R.G. and Torralba, A. Low supply voltage high-performance CMOS current mirror with low input and output voltage requirements. *IEEE Transactions on Circuits and Systems II: Express Briefs*, vol. 51, no. 3, pp. 124-129, March 2004.

[25] C. Muñíz. *Corrección de Offset en Líneas de Retardo de Frecuencia Intermedia*. M Sci. thesis, INAOE, Tonanzintla, Puebla, México, 2003.

PART III: PERFORMANCES

In: Analog Circuits: Applications, Design and Performance
Editor: Esteban Tlelo-Cuautle

ISBN: 978-1-61324-355-8
© 2012 Nova Science Publishers, Inc.

Chapter 7

GENERALIZED APPROACH FOR ANALOG NETWORK OPTIMIZATION

A.M. Zemliak

Autonomous University of Puebla, Department of Physics and Mathematics,
Av. San Claudio s/n, Ciudad Universitaria, Puebla. 72570, MEXICO.
National Technical University of Ukraine "KPI", Institute of Technical Physics,
Av. Peremogy 37, Kiev, 03056, UKRAINE

Abstract

The problem of reducing the computing time of design of large systems is one of the pressing problems of the general improvement of the quality of designing. This problem is of particular importance for VLSI of electronic circuits and it was discussed on the basis of the new approach.

The methodology of analog system design includes two main parts: the unit of analysis of mathematical models of the system and block parameter optimization, which achieves the optimum point of the objective function. There are some powerful techniques that reduce the time required for the analysis of electronic circuits: the idea of using the structure of sparse matrices, the methods of decomposition, and macro model representation. Advances in optimization's technique also help to develop the fast algorithms for the design of electronic circuits. Nevertheless, the analysis of large integrated circuits and the time required for the optimization procedure increases when increasing the size and complexity of the circuit. Meanwhile, it is possible to reformulate the general problem of designing an electronic circuit and to generalize the process of optimization.

A generalized approach to optimization of analog circuits can be submitted on the basis of control theory. First of all, this approach serves to minimize the time of the circuit design. On the other hand, this approach makes it possible to analyze with great clarity the optimization process when moving along the trajectory in phase space of designing process. The basic concept of this technique is the introduction of a special control vector, which on the one hand, generalizes the design process and, on the other hand, it allows you to control the optimization process to achieve the optimum objective function of the minimum CPU time. This control vector serves as the primary instrument for the redistribution of computational cost between the procedure of analysis and optimization procedure. Opportunity to minimize the CPU time is due to an infinite number of different design strategies that exist in the framework of the generalized theory. This idea can be incorporated into any optimization method.

The problem of searching the strategy of designing at the minimal possible time is formulated now as a typical problem of minimizing a functional of the optimal control theory. The process of optimizing the system is formulated in this case, as controlled dynamic process. The main objective of this process is to minimize the computer time using the optimal choice of the control vector during optimization.

The problem of finding an optimal control vector can be solved on the basis of Lyapunov's direct method. The behavior of the Lyapunov function of dynamic process and its time derivative are the sufficient information to select the best design strategy of an infinite number of different strategies.

Introduction

One of the main problems when designing large systems is the task of reducing the time taken to achieve the optimum point of the objective function of the design process. The design process includes optimizing the structure of the future system, but as this stage is connected with the unsolved problem of artificial intelligence, then, in general, it is done manually and, therefore, not in the system CAD. In other words, the traditional approach to computer-aided design of electronic circuits consists of two main parts. First the analysis model of the electronic circuit, described by a system of algebraic or integro-differential equations, and secondly from the procedure of parametric optimization, which leads to find the optimum objective function that meets all specifications. There are some powerful techniques that reduce the time required for circuit analysis. Because the matrix of large-scale schemes are very sparse, successfully used special mathematical techniques for sparse matrices [1, 2]. Another approach to reduce the number of computations required for solving linear and nonlinear equations, based on the methods of decomposition. The partitioning of a circuit matrix into bordered-block diagonal form can be done by branches tearing as in [3], or by nodes tearing as in [4] and jointly with direct solution algorithms gives the solution of the problem. The extension of the direct solution methods can be obtained by hierarchical decomposition and macromodel representation [5]. Other approach for achieving decomposition at the nonlinear level consists on special iteration techniques and has been realized for example in [6-8] for the iterated timing analysis and circuit simulation. A technique of optimization that is used to optimize the circuit has a very strong impact on the overall computer time too. Computational methods are developed for both the unconstrained and constrained optimization [9, 10]. Practical aspects of using these methods are developed for VLSI design, and optimization of yield, time and area [11-13]. It can be assumed that the methods of circuit analysis and optimization procedures will be improved in the future. Meanwhile, it is possible to reformulate the general problem of designing and generalize it for a variety of different design strategies. Clearly, of course, that a large number of different strategies include more opportunities to select one or several design strategies that are optimal or quasi-optimal in time. This is especially true if we have an infinite number of different design strategies. The time required for optimization is growing rapidly with increasing system complexity. Using known methods of reducing the time to analyze the system in the traditional approach were insufficient. We say that the conventional approach to the design of electronic circuits the traditional design strategy, in the sense that the analysis method based on Kirchhoff's law. A new formulation of the circuit optimization without strict compliance with the laws of Kirchhoff has been proposed [14, 15]. This process was called the generalized optimization and used the idea of ignoring the Kirchhoff laws for the whole circuit or any part thereof. In this case, in addition to minimizing a predefined objective function, we also need to minimize the penalty function, which includes residual equations of the model circuit. In the extreme case, when the penalty function includes all equations of the model circuit, this idea has been practically implemented in two systems CAD [16-18]. The latter idea can be called a modified traditional design strategy. In contrast to the traditional strategy, when the analysis model circuit is carried out at each step of the optimization procedure, a modified traditional strategy can be defined as a strategy that does not include the analysis of the model in the optimization process. Another formulation of the problem of designing circuits based on the idea of generalizing the design process leads to a number of different design strategies. In this case, we can formulate the problem of finding a strategy that allows solving the problem of optimization circuit for minimum CPU time. Then the optimal strategy design can be defined as a strategy to achieve the optimum point of the objective function with minimal downtime. The main problem with this definition is to identify the conditions necessary for constructing an optimal algorithm. The solution to this problem would substantially reduce the computing time required for circuit design.

Schematic Design as Optimization Problem

Although the task of designing electronic circuits is a typical nonlinear programming problem, in practice, the design problem is solved in another way. In the presence of M constraint equations $g_i(X)=0$, $i=1,...,M$, any K variables ($K=N-M$) of N variables $X = (x_1,..., x_N)$ can be chosen independently of each other, and the rest of M coordinates will be determined by the system of equations $g(X)=0$. On the role of the independent variables most often selected circuit elements such as resistors, capacitors, inductors, etc. At the same time as the dependent variable is natural to define currents and voltages in the circuit.

Supposed to be given the specification circuit, i. e. all the required characteristics of the designed object. Based on the experience of the developer and additional information available concerning possible variants of the system being developed, selected a suitable circuit topology. This stage of the design has not yet subject to full automation because of the unsolved problems of artificial intelligence. Automated part of the system design begins with the job circuit topology, and thus, all other components of the system design can be algorithmizing. The traditional process of parametric circuit optimization in a mathematically formalized aspect is shown in Figure 1.

The choice of initial values of the circuit components corresponds to the choice of the initial point in the space of independent state variables, i. e., the zero step of the iterative procedure are given initial values of independent variables $X'^s = (x_1^s, x_2^s,..., x_K^s)$ for $s = 0$. The analysis of the object is provided after (circuit, device, system), which means the analysis of mathematical models to find the dependent variables, i. e., solve the system $g(X)=0$. In the mathematical sense this part corresponds to the implementation constraints in the problem of constrained optimization and, thus, determines the values of missing variables in the system that are dependent. Based on the knowledge of all the variables of the system, the value of the objective function $C(X^s)$ is calculated for the s-th iteration step and the condition of achieving the minimum of this function is checked. In the case of reaching the minimum of target function, the solution of the design problem is concluded with the desired characteristics for the designed circuit. If no minimum of the objective function, the general design algorithm be continue with the optimization which defines the direction of motion in the space of independent variables H^s and modifies the values of independent variables for the $s+1$ iteration step. Then the whole process is repeated. The number of repetitions (iterations of the optimization procedure) depends on the optimization method, the successful choice of initial approximation for the independent variables of the system, and the necessary accuracy of the solution.

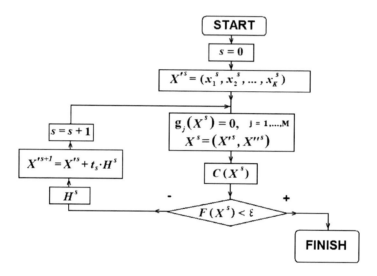

Figure 1. The traditional process of parametric optimization of circuit.

Problem of designing an electronic system can be reformulated, following the idea [14], by means of generalizing and formalizing it for a variety of different design strategies. In this case, you can put the task of selecting from this set, one strategy, the best in some sense, for example, in the sense of speed. In terms of CPU time, the optimal strategy design can be defined as a strategy that achieves the optimum point of the objective function of the design process in minimum time. In this case, you can put the task of selecting from this set, one strategy, the best in some sense, for example, in the sense of speed. The answer to this question will significantly reduce the cost of computer time required for design purposes.

The idea of using optimal control theory for the generalization of the design of electronic systems has been proposed in several papers [19-24] and described in detail below. Depending on the presentation of the optimization procedure, the problem of designing analog systems can be formulated in discrete or continuous formulation. Discrete formulation of the more accepted it is in the design of electronic circuits, while the continuous formulation is more typical of optimal control problems. In the latter case, the continuous form of equations of the optimization is more adequate for the application of different methods of optimal control theory. In fact, there is no significant difference between these two forms and one passes into another when appropriate limit.

In general, as will be shown, the formalization of design problem based on the ideas of control theory generalizes the problem of designing and significantly reduces the required computer time for system design in the case of constructing an optimal algorithm.

Formulation of the problem in a discrete form. From a mathematical point of view, following [23], we call the traditional strategy of designing the problem of designing analog systems on the basis of the given topology by means of constrained minimization of the objective function $C(X)$ in space R^N, where N is the total number of variables of electronic circuit and the vector X includes all variables of the circuit. The optimization process can be defined for example in the form of two-step procedure, the following:

$$X^{s+1} = X^s + t_s \cdot H^s, \tag{1}$$

where s is the iteration number of the optimization procedure, t_s the iterative parameter, $t_s \in R^1$ the function H determines the direction of movement in space variables R^N and its form is determined by one or another method of minimizing the objective function $C(X)$.

It is understood that the objective function $C(X)$ is the criterion for the design and contains all the design objectives, i. e. the achievement a minimum of this function leads to the attainment of design objectives. The limitations of this procedure of constrained optimization are presented as the system of equilibrium equations of electronic circuits, which can be formulated in the form of Kirchhoff's equations, the method of nodal potentials, the method of contour currents, etc. To perform some physical relations and properties can be determined by additional constraints in the form of inequalities. We assume further that the system of constraints in the form of equations, i. e. mathematical model of the electronic circuit can be described by a system of nonlinear algebraic equations:

$$g_j(X) = 0, \qquad j = 1, 2, \ldots, M \tag{2}$$

for M dependent components of the vector X.

This same traditional strategy of design can be determined using the procedure of unconditional minimization of the objective function $C(X)$ in the space R^K where K is the number of independent variables, while simultaneously solving the system (2) for M dependent components of the vector X. Vector $X \in R^N$ is divided into two parts: $X = (X', X'')$, where the vector $X' \in R^K$ is a vector of independent variables, the vector $X'' \in R^M$ is a vector of dependent variables and $N=K+M$. It is clear that this division to the dependent and independent variables is quite arbitrary, because any parameter can be taken as an independent or dependent. With this definition, some of the parameters of the design process as a frequency, temperature, etc., remain outside consideration. But there is no impediment to including them in the general procedure for the design process, and in this case it can be assumed constant and their attributed to the coefficients of system (2).

The process of minimizing the objective function $C(X)$ in the space R^K of independent variables for the two-step optimization procedure is generally described by the following vector equation:

$$X'^{s+1} = X'^{s} + t_s \cdot H^s \tag{3}$$

Constraints of the independent variables that may arise from their physical nature is easily bypassed, and these be done when considering specific examples of designing different circuits. Described the traditional approach to design includes the analysis of electronic circuits, i. e. solution of the nonlinear system (2) at each step of the optimization procedure. This traditional approach we call as a traditional strategy of design (TSD). The interaction of various elements of TSD is shown on the block diagram in Figure 1.

In the case of designing a VLSI, the system (2) has a very high order, which is one of the main reasons for the increase of computer design time. At the same time, the concept of the traditional strategy of design does not exclude, but rather involves the use of advanced optimization algorithms and new ideas aimed at reducing the analysis time of large systems.

Specificity of the process of designing systems, at least electronic circuits, is that there is no need to comply with the conditions (2) for each step of the optimization process and it is quite sufficient in this case to satisfy the conditions (2) in the end point of the design process. A heuristic similar idea for the design of electronic circuits was proposed in [14-15]. With this approach, the vector function of the procedure of optimization H depends not only on the objective function $C(X)$, but also additional penalty function $\varphi(X)$, structure of which should include all equations (2). Formalizing this idea, primarily involves the problem of definition of special form of penalty function, which can be defined like this:

$$\varphi(X^s) = \frac{1}{\varepsilon} \sum_{j=1}^{M} g_j^2(X^s) \tag{4}$$

where the parameter ε is introduced for further adjustment. In many cases, this parameter can be set equal to 1. In this case, we define the design process as a problem of unconstrained optimization (5):

$$X^{s+1} = X^s + t_s \cdot H^s \tag{5}$$

in space R^N without any additional system constraints, but for the new target function $F(X)$, which should include information on the electronic circuit in the form of an additional penalty function (4) and can be determined an additive expression (6).

$$F(X) = C(X) + \varphi(X) \tag{6}$$

With this approach, when the minimum objective function $F(X)$ is achieved, at the same time we achieve the minimum objective function $C(X)$ and satisfy the system (2) at the end point of the optimization process. At the same time we can say that throughout the entire process of designing the laws of Kirchhoff's circuits are not satisfied at any step of the optimization procedure. This method can be called as a modified traditional method of design, and it reproduces the other strategy of design that we call the modified traditional strategy of design (MTSD), which has other trajectory in space R^N. In this case, as we see, the analysis of electronic circuits, i. e., the solution of (2) is completely absent, but all the complexity of the problem is transferred to the solution of the optimization procedure. The optimization process must be carried out in space R^N, but not in space R^K as in the case of the traditional design strategy, and the function $F(X)$ is much more complex than the original function $C(X)$. The block diagram of a modified traditional strategy of designing is presented in Figure 2.

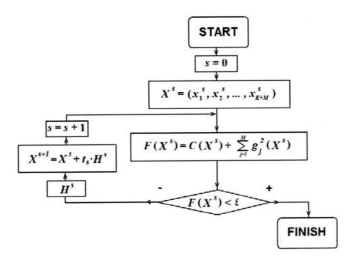

Figure 2. The algorithm of the modified traditional strategy of designing.

As is evident from the substance of the proposed MTSD, in this case there are no costs of computer time to analyze the circuit, but increase the cost of optimization procedure, since the space R^N has a greater dimension than R^K. Priori not known which of these two strategies has less computer time and can we get a win in the CPU time when using the MTSD? This issue requires further study.

Formulation of the problem in a continuous manner. Specificity of the formulation of the problem of design in this case lies in the fact that the optimization procedure instead of equations (1) or (5) is written as a system of ordinary differential equations:

$$\frac{dX}{dt} = f(X) \qquad (7)$$

where the right hand side $f(X)$ of the system is determined by the method of optimization and essentially coincides with the previously introduced function H. It is clear that the discretization of equations (7) lead us to the system (1) or (5). Nevertheless, for purposes of analysis, shape of optimization's procedure, given by (7) is convenient. For the further generalization of the design problem will be shown that the continuous form of the optimization procedure is more natural in terms of control theory.

A number of different design strategies. Structural basis of the process of designing. The idea of applying the additional penalty function can be extended if the penalty function includes only part of the system (2), while the rest of the system (2) is interpreted as a system of restrictions. In this case, the penalty function includes, for example, only Z first members:

$$\varphi\left(X^{s}\right) = \frac{1}{\varepsilon} \sum_{i=1}^{Z} g_i^2\left(X^{s}\right) \tag{8}$$

where $Z \in [0, M]$, while the remaining $M\text{-}Z$ equations giving the modified system of equations (9), which obtained from the system (2):

$$g_j(X) = 0, \quad j = Z+1, Z+2, ..., M \tag{9}$$

It is quite clear that each new value of number Z produces a new strategy of designing and a new trajectory in parameter space. In other words, each strategy of design is determined by its own system of constraints (9) and its own optimization procedure in which the objective function (6) includes a penalty function (8), including, in turn, Z first equations of the original system (2).

This idea can easily be generalized to the case when the penalty function $\varphi(X)$ includes Z arbitrary equations of the system (2). In this case, the total number of different design strategies is equal 2^M. All of these strategies exist within the same optimization procedure. Optimization's procedure is realized in the space R^{K+Z} and the number of equations that need to be solving at each step of the optimization procedure is equal to $M\text{-}Z$. The total set of strategies that appear in this approach creates a generalized strategy of design. Number of dependent parameters M increases with the complexity of the system, in this case the electronic circuit, and can reach hundreds of thousands and millions for VLSI. It is clear that the number of different design strategies increases exponentially as 2^M, and their number is huge, but nonetheless finite. All these strategies have the same starting point S in parameter space, since the choice of starting point does not depend on the strategy and they have the same final point F (Figure 3), because the end point defines the result of the design process. It is understood that this is a unique solution. The case of ambiguity solutions discussed further in the analysis of examples.

What is the difference between all the strategies of this set? All these strategies have a different number of operations, a different total processing time and the various paths in parameter space. In this case it is possible to formulate the task of finding the strategy of design optimal in time, which has the minimum number of operations and CPU time. Here and below the optimality of the design process we understand in the sense of a minimum of CPU time. The total number of different strategies in a generalized strategy of design that reaches values 2^M is called a structural basis for the design.

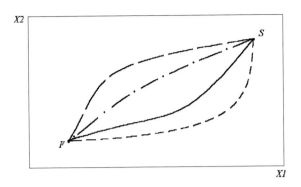

Figure 3. Various possible trajectories in design space parameters.

Traditional strategy of designing (TSD) includes two systems of equations. Let us consider for concreteness that the optimization procedure is based on the gradient method and can be determined through a system of ordinary differential equations for the independent variables as follows:

$$\frac{dx_i}{dt} = -b \cdot \frac{\delta}{\delta x_i} C(X), \qquad i = 1, 2, \ldots, K \qquad (10)$$

where b is the iteration parameter. Operator $\delta / \delta x_i$ here and below means:

$$\frac{\delta}{\delta x_i} \sigma(X) = \frac{\partial \sigma(X)}{\partial x_i} + \sum_{p=K+1}^{K+M} \frac{\partial \sigma(X)}{\partial x_p} \frac{\partial x_p}{\partial x_i} \qquad (11)$$

Using the method of gradient does not diminish the generality of obtained results. For any other method you only need to define the process of circuit optimization as a system of ordinary differential equations for the independent variables.

Modified traditional strategy of designing (MTSD) is completely defined by means of the procedure of optimization without any additional restrictions. That is the system (2) is absent. The number of independent variables is equal to $K+M$. The basic system is as follows:

$$\frac{dx_i}{dt} = -b \cdot \frac{\delta}{\delta x_i} F(X), \qquad i = 1, 2, \ldots, K + M \qquad (12)$$

Fanction $F(X)$ is the generalized objective function and is defined as:

$$F(X) = C(X) + \frac{1}{\varepsilon} \sum_{j=1}^{M} g_j^2(X) \qquad (13)$$

A generalized strategy of design has a variable number of independent parameters, which is equal to $K + Z$. In this case, the following two equations are used:

$$\frac{dx_i}{dt} = -b \cdot \frac{\delta}{\delta x_i} F(X), \qquad\qquad i = 1,2,\ldots, K + Z \qquad\qquad (14)$$

$$g_j(X) = 0, \qquad\qquad j = Z + 1, Z + 2,\ldots, M \qquad\qquad (15)$$

where
$$F(X) = C(X) + \frac{1}{\varepsilon} \sum_{j=1}^{Z} g_j^2(X).$$

It is clear that when the design of VLSI among the vast number of possible design strategies that exist in limits of generalized approach should be the strategy that has less CPU time than the traditional strategy of design.

Formalization of the design process circuits based on control theory

The most general approach to the problem of constructing optimal algorithm of design can be developed through the application of optimal control theory. It is possible to compile a more generalized strategy of design, than defined above, on the basis of equations (7)-(9) if we assume the possibility of changing the parameter Z during the optimization process. This means that we can change the number of independent variables, $K+Z$ and number of members in the penalty function formula (8) Z at each point of the procedure of optimization. To implement this idea, it is convenient to introduce the vector of control functions or control's vector $U = (u_1, u_2, \ldots, u_m)$, where $u_j \in \Omega$, $j = 1, 2, \ldots, M$, $\Omega = \{0;1\}$. That is, each control function u_j can only take one of two values 0 or 1. These functions have the sense of control functions of the design process as they allow reallocating computational complexity, i. e. the number of operations between the optimization's procedure (3) or (7) and analysis of mathematical models of the circuit (9). This control's vector generalizes the design process. The meaning of the control function u_j is as follows: the equation number j belongs to the system (9), while a member $g_j^2(X)$ is removed from the right-hand side of (8) when $u_j = 0$, and vice versa, the equation j is removed from the system (9), while the corresponding term appears in the right side (8) when $u_j = 1$. In this case, the mathematical model of the electronic circuit can be written as follows:

$$(1 - u_j) g_j(X) = 0, \quad j = 1,2,\ldots, M \qquad\qquad (16)$$

The formula that determines the type of penalty function is changed to the following:

$$\varphi(X) = \frac{1}{\varepsilon} \sum_{j=1}^{M} u_j \cdot g_j^2(X) \qquad\qquad (17)$$

All control functions u_j are the functions of the current point of the optimization process. In this case, the vector of directed motion H is a function of the vectors X and U:

$H = f(X,U)$. The physical meaning imposed control functions is the redistribution of computing time between the block of analysis, the system (16), and the block of optimization, which includes a penalty function (17).

Given the continuous form of the optimization's procedure can be said that the total number of different design strategies, that produced inside the same optimization procedure is virtually unlimited. Among all these strategies, there is one or more strategies that are optimal and they can achieve all the objectives of designing for the fastest possible time. Hence the problem of finding the optimal design strategy is formulated now as a typical problem of minimizing a certain functional in the theory of optimal control and reduces to the search of optimal control. The functional is defined as a real computer time required for the solving of the design problems. As the main instrument is the vector of control functions U. The main problem in this definition is a searching of optimal dependency of control's vector. It is clear however, that if an optimal vector of control functions Uopt available, and then the optimal strategy for the circuit's design is implemented using this vector.

The idea of the formulation of the problem of system design in optimal time as a problem of functional minimization in optimal control theory does not depend on the specific implementation of the optimization algorithm and can be embedded in any optimization procedure. This is shown in [21-24], which tested three different algorithms, which are typical representatives of three major groups of optimization methods: gradient method, Newton's method and the method of Davidon-Fletcher-Powell (DFP) [9].

In the above formulation, any possible strategy of design is defined by means of the vector U, and has its own trajectory in the space of variables. It is clear however that the comparison of different trajectories of movement by time or on any other parameter adequately only if these paths have the same initial and final points. On the other hand, the objective function of the design process C (X) can have many local minima, because the task of designing, being the inverse problem with respect to the problem of analysis and it is nonlinear, even if we design the physically linear system that described by a linear mathematical model. In this case, to adequately compare different strategies and their trajectories is desirable to impose additional conditions for the uniqueness in order to achieve one and the same the end point in parameter space. At the same time, the problem of no uniqueness is not a specific feature of the new formulation of design methodology. We have a problem of this type, whenever begin the design process from different starting points.

Continuous form of the design process. The process of designing the system in the formulation of control theory can be written in a discrete and continuous form. Continuous form is adequate more in the spirit of optimal control theory. To represent the problem in a continuous manner, you can use a vector differential equation (7), which is modified by the introduction of the control's vector U:

$$\frac{dX}{dt} = f(X,U) \tag{18}$$

where the right hand side $f(X,U)$ is a vector of directional movement of H and depends on a generalized objective function $F(X,U)$ defined, for example by the following expression:

$$F(X,U) = C(X) + \frac{1}{\varepsilon}\sum_{j=1}^{M} u_j g_j^2(X) \tag{19}$$

This means that the design process is formulated as a problem of integrating the system (18) with the additional conditions (16). Structure of the function $f(X,U)$ can be defined as follows:

$$f(F(X,U)) = -F'(X,U) \tag{20}$$

for the gradient method,

$$f(F(X,U)) = -\{F''(X,U)\}^{-1} \cdot F'(X,U) \tag{20'}$$

for the Newton's method, where $F''(X,U)$ is the matrix of the second derivatives,

$$f(F(X,U)) = -B(X,U) \cdot F'(X,U) \tag{20''}$$

for the DFP method, where $B(X,U)$ is a symmetric, positive definite matrix for DFP algorithm.

In this case, the problem of designing an optimal in time algorithm of design is formulated as a typical problem of minimizing the functional of the control theory for systems of differential equations (18) with right hand part, depending on the type of a particular method of optimization, such as (20) (20 ') or (20"), with the objective function, which defined by (19) and constraints (16), which is defined as a modified mathematical model of the circuit. An additional complication is that the right hand side of (18) are not continuous functions but only piecewise continuous. A similar problem for the system (18) with piecewise continuous control functions can be solved most adequately by the well-known method of Pontryagin's maximum principle [25], but the direct application of this principle for the nonlinear problem of large dimension is very problematic. This problem can be solved adequately for linear systems only [26]. To solve this problem can be approached on the basis of ideas developed in the course of an approximate solution of problems of control theory [27-31].

The basic system of equations in the process of design for the three previously mentioned optimization algorithms are presented below. System (18) can be written component-wise form as follows:

$$\frac{dx_i}{dt} = f_i(X,U), \quad i = 1,2,\dots, K, K+1,\dots N \tag{21}$$

This system together with the system (3.1) defines the design process.

In the case of the gradient method for the right-hand side of equation (21) has the form:

$$f_i(X,U) = -\frac{\delta}{\delta x_i} F(X,U), \qquad i = 1,2,...,K \qquad (22)$$

$$f_i(X,U) = -u_{i-K} \frac{\delta}{\delta x_i} F(X,U) + \frac{(1-u_{i-K})}{dt} \left[-x_i' + \eta_i(X) \right], \quad i = K+1, K+2,..., N \qquad (22')$$

where x_i^s is equal to $x_i(t-dt)$ and the function $\eta_i(X)$ written in implicit form, determines the current value of x_i^{s+1}, $\left(x_i^{s+1} = \eta_i(X) \right)$, which is a result of solving system (16). Variables of control u_j are functions of the "current time" of optimization process. Systems (16)-(21) define the design process in a continuous manner. The formulas (21) and (22) define the process of evolution for the independent variables. The formulas (21) (22') define the variables that were initially identified as dependent variables, but can also be transformed into independent. The equation with number j disappears from the system (16) and the dependent variable x_{K+j} is transformed into an independent, when $u_j = 1$. This variable is defined in such a case, by (21) (22'). In this case, there is no difference between the formulas (22) and (22') because variable x_{K+j} is an ordinary independent variable. On the other hand, equation (3.6) with right part (22') is transformed into an identity $\frac{dx_i}{dt} = \frac{dx_i}{dt}$ when $u_j = 0$ because $\eta_i(X) - x_i' = x_i(t) - x_i(t-dt) = dx_i$. This means that the variable x_i is a dependent variable at this moment and its current value is determined directly from the system (16). The transformation of vectors X' and X'' can be made at any time. This leads to a virtually unlimited number of different design trajectories generated by the system (16)-(21). Each trajectory corresponds to the own strategy.

In the case of Newton's method or the method of DFP the equations (22) and (22') mutate and acquire the following form:

$$f_i(X,U) = -\sum_{k=1}^{N} b_{ik} \frac{\delta}{\delta x_k} F(X,U), \qquad i = 1,2,...,K \qquad (23)$$

$$f_i(X,U) = -u_{i-K} \sum_{k=1}^{N} b_{ik} \frac{\delta}{\delta x_k} F(X,U) + \frac{(1-u_{i-K})}{t_s} \left\{ -x_i^s + \eta_i(X) \right\}, \quad i = K+1, K+2,...,N \qquad (23')$$

where b_{ik} is the element of the matrix of second derivatives $\{F''(X,U)\}^{-1}$ for Newton's method or element of the matrix $B(X, U)$ for the method of the DFP. In the latter case, the matrix $B(X, U)$ is defined by the following expressions [9]:

$$B_{s+1} = B_s + \frac{R^s (R^s)^T}{(R^s)^T Q^s} - \frac{(B_s Q^s)(B_s Q^s)^T}{(Q^s)^T B_s Q^s}, \qquad (24)$$

where B_0 is a unitary matrix $s = 0,1,...$ and $R^s = X^{s+1} - X^s$, $Q^s = F(X^{s+1}, U^{s+1}) - F(X^s, U^s)$.

Numerical results corresponding to the new, generalized approach to the formulation of the design process are presented below. These results demonstrate the potential opportunities in the construction of an optimal algorithm in time.

Discrete form of the design process. This form of presentation of the design process differs from previous ones in that instead of a system of differential equations (18) the process of optimization described by the system, which is modified from the system (5). The modification is to include a vector of control functions as an additional variable function H, which defines the direction of reducing the objective function $C(X)$ or $F(X)$. In vector form, this system has the form:

$$X^{s+1} = X^s + t_s \cdot H^s(X,U) \qquad (25)$$

Componentwise, this system can be rewritten as:

$$x_i^{s+1} = x_i^s + t_s \cdot f_i(X,U), \quad i = 1,2,..., K, K+1,...N \qquad (26)$$

where functions $f_i(X,U)$ are determined above.

In reality, the difference between the continuous form of the design process and the discrete form is purely philosophical, because the continuous form is a system of differential equations, and it is solved by numerical methods, i. e. by the discrete manner. At the same time as the system (21) and system (25) can be solved with a constant step of integration, and with variable pitch, selected on the basis of one or another concept of optimal steps. In further analysis, we used both forms for the process of optimization, both the discrete and continuous.

A generalized formulation of the circuit design process, given in the preceding paragraph, based on optimal control theory, lets you examine and compare different design strategies by modifying the control vector U. Let us analyze some examples of the design of electronic circuits, both passive and active. Optimization of the circuit is accomplished by three different methods, the gradient method, Newton's method and the method of Davidon-Fletcher-Powell (DFP), which are basic for the three groups of methods of optimization - the first order, second order, and one and a half order of magnitude. Consider the problem of finding an optimal design strategy for a simple non-linear passive electronic circuit shown in Figure 4. This circuit has two nodes and is characterized by three independent variables ($K = 3$) and two dependent ($M = 2$). In this case, the total number of variables N is 5.

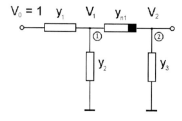

Figure 4. The scheme of nonlinear passive electronic circuit with two nodes.

Vector X is defined by five components, which are given by: $x_1^2 = y_1$, $x_2^2 = y_2$, $x_3^2 = y_3$, $x_4 = V_1$, $x_5 = V_2$. Determination of the components x_1, x_2, x_3 of the proposed formulas will automatically lead to positive values of conductivity, which eliminates the problem of positive definiteness of the resistance and conductivity, and allows optimization on the entire space of values of these variables without any restrictions. A mathematical model of the circuit is as follows:

$$(1 - u_1)g_1(X) = 0$$

$$(1 - u_2)g_2(X) = 0$$

(27)

The generalized objective function $F(X,U)$ of the optimization process is given by the following expression:

$$F(X,U) = C(X) + \frac{1}{\varepsilon}\left(u_1 g_1^2(X) + u_2 g_2^2(X)\right)$$

(28)

where the objective function is defined by $C(X) = (x_5 - k_V)^2$, where k_V has a fixed value $k_V = V_{out}/V_0 = V_{out}$.

Process of optimization in continuous form is defined by the system (21), where the right part, for example for a gradient optimization method defined by the following expressions:

$$f_i(X,U) = -\frac{\delta}{\delta x_i}F(X,U), \quad i = 1,2,3$$

(29)

$$f_i(X,U) = -u_{i-K}\frac{\delta}{\delta x_i}F(X,U) + \frac{(1 - u_{i-K})}{dt}\left[-x_i' + \eta_i(X)\right], \quad i = 4,5$$

(29′)

The results of the analysis of different strategies for the complete structural basis and for the three optimization methods are presented in Table I, provided the integration of system (21) with a variable optimal step.

Table I. Results of the analysis of full structural basis of different design strategies for passive nonlinear circuit with two nodes

N	Control vector U (u_1, u_2)	Gradient method		Newton's method		DFP method	
		Iterations number	Time (sec)	Iterations number	Time (sec)	Iterations number	Time (sec)
1	(0 0)	16	0.0243	7	0.0396	8	0.0241
2	(0 1)	51	0.0238	9	0.0251	10	0.0127
3	(1 0)	60	0.0448	8	0.0329	21	0.0331
4	(1 1)	68	0.0217	11	0.0231	23	0.0198

The first row of the table, the strategy 1 corresponds to the TSD while the control vector U, which consists of two components, equal to (00) and at each step of the iterative procedure is solved a system (27). Optimization procedure is carried out in space R^3 defined by the system (21). The last row of the table, the strategy of 4 corresponds to the MTSD, and the control's vector is equal to (11). This means that the system (27) disappears, and the optimization procedure is carried out in space R^5 defined by the system (21) for $i = 1,2,3,4,5$ with right-hand part of (29) and (29'). In this case, the form (29) and (29') coincides, as both control functions u_1 and u_2 are equal to unity. The generalized objective function in this case includes two additional terms in (28). The remaining two strategies for the structural basis are intermediate. Strategy 2 is characterized by the vector control (01), which means no second equation (27) and the appearance of the second additional term in (28). Optimization's procedure is carried out in space R^4 defined by the system (21) for $i = 1,2,3,5$. Finally, strategy 3 is characterized by the vector control (10), which means no first equation (27) and the appearance of the first additional term in (28). Optimization procedure is carried out in space R^4 defined by the system (21) for $i = 1,2,3,4$.

Comparison of all four strategies shows that strategy 4, i. e., MTSD is optimal for the gradient method and the Newton's method. In this case, winning in a computer time in comparison with the TSD is 1.12 times and 1.71 times respectively. At the same time, the DFP method has the intermediate strategy 2 with control's vector (01) as the best strategy. At the same time the time gain compared to the TSD was 1.9 times. However, it is clear that these strategies, being among the best strategies for the structural basis, i. e. with a constant vector of control, are not optimal in general. Optimal strategy was searched among the set of strategies with varying control's vector, i.e., when the transition from one strategy to another during the optimization process. Optimal strategy, or rather quasi-optimal one, was found for all three optimization methods due to the special optimization procedure with variations of the control vector. The results of this additional optimization in the case of a single switch from one strategy to another are presented in Table II.

Table II. Results of the analysis of quasi-optimal design strategies for passive nonlinear circuit with two nodes

N	Method	Optimal control vector U (u_1, u_2)	Iterations number	Switching point	CPU time (sec)
1	Gradien	(1 0); (1 1)	39	11	0.0161
2	Newton	(1 1); (1 0)	7	3	0.0218
3	DFP	(0 0); (1 1)	10	9	0.0115

The obtained quasi-optimal strategies lead to a reduction in computing time in comparison with TSD and was obtained by the gain in 1.5 times, 1.82 times and 2.09 times respectively for the gradient method, Newton's method and the method of the DFP. Clearly, that this gain in time is not significant, however, as will be shown below, this gain increases many times when electronic circuit is more complex.

Figure 5 presents a nonlinear electronic circuit with seven nodes, representing a three-stage transistor amplifier. This circuit is determined by seven independent variables $y_1, y_2, y_3, y_4, y_5, y_6, y_7$ ($K = 7$) and seven dependent variables $V_1, V_2, V_3, V_4, V_5, V_6, V_7$ ($M = 7$). Vector X has 14 components, and they are defined by the following formulas:

$$x_1^2 = y_1, \quad x_2^2 = y_2, \quad x_3^2 = y_3, \quad x_4^2 = y_4, \quad x_5^2 = y_5, \quad x_6^2 = y_6, \quad x_7^2 = y_7, \quad x_8 = V_1, \quad x_9 = V_2, \quad x_{10} = V_3,$$
$$x_{11} = V_4, \quad x_{12} = V_5, \quad x_{13} = V_6, \quad x_{14} = V_7.$$

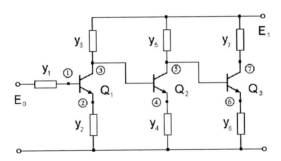

Figure 5. Three-stage transistor amplifier.

Sources E_0, E_1 are given. As a model of the transistor selected known Ebers-Moll model of the DC used in the SPICE [32]. The purpose of designing this circuit is the correct choice of conductivities, which provides the required voltages on the emitter and collector V_{EB0} V_{KB0} junctions of all transistors. The objective function is defined as the sum of squared differences between the preset and current values of the voltage across the junction transistor.

In this case, the system optimization procedure (21) consists of 14 equations, and the model of circuit is determined by the 7 nonlinear equations. Structural basis of strategies of design has 128 different strategies. Results of the analysis of traditional strategy of design and some strategies of the structural basis, having lesser computer time than traditional strategy for variable optimal step integration of the system (21), presented in Table III.

Table III. Results of the analysis of the full structural basis of different design strategies for the three-stage transistor amplifier with a variable integration step

N	Vector of control functions U (u1,u2,u3,u4,u5,u6,u7)	Gradient method Iterations number	Gradient method Time (sec)	DFP method Iterations number	DFP method Time (sec)
1	(0 0 0 0 0 0 0)	6379	321.09	854	64.47
2	(0 0 1 0 1 0 1)	922	54.53	764	52.29
3	(0 0 1 0 1 1 0)	1667	80.71	650	46.13
4	(0 0 1 0 1 1 1)	767	35.35	426	22.68
5	(0 0 1 1 1 0 0)	3024	159.67	940	52.71
6	(0 0 1 1 1 0 1)	823	37.73	177	7.71
7	(0 0 1 1 1 1 0)	3068	86.87	450	14.56
8	(0 0 1 1 1 1 1)	553	15.75	170	6.93
9	(0 1 1 0 1 0 1)	465	10.01	101	2.66
10	(0 1 1 0 1 1 0)	1157	31.92	111	3.85
11	(0 1 1 0 1 1 1)	501	8.82	124	2.66
12	(0 1 1 1 1 0 0)	2643	72.66	314	9.24
13	(0 1 1 1 1 0 1)	507	9.24	170	4.62
14	(0 1 1 1 1 1 0)	3070	67.27	423	12.25
15	(1 0 1 0 1 0 1)	1345	28.07	397	16.94
16	(1 0 1 0 1 1 1)	615	10.01	191	4.62
17	(1 0 1 1 1 0 1)	699	10.71	197	4.97
18	(1 0 1 1 1 1 1)	366	4.97	103	1.96
19	(1 1 1 0 1 0 1)	789	10.43	201	4.97
20	(1 1 1 0 1 1 0)	3893	61.53	1158	18.06
21	(1 1 1 0 1 1 1)	749	7.71	148	2.11
22	(1 1 1 1 1 0 0)	4325	90.72	945	19.18
23	(1 1 1 1 1 0 1)	796	8.47	133	2.31
24	(1 1 1 1 1 1 0)	2149	29.26	1104	13.44
25	(1 1 1 1 1 1 1)	2031	5.67	180	0.77

Among all the strategies of this table, the strategy 18 with the vector of control (1011111) is optimal for the gradient method and strategy 25, corresponding to the MTSD with the control's vector (1111111) is optimal for the method of the DFP. They allow us to obtain a gain in time compared to traditional to 64.6 times for the gradient method, and 83.7 times for the method of the DFP. However, as in previous examples, the quasi-optimal strategies for both methods were found by means of a special optimization procedure. The corresponding results are presented in Table IV.

Quasi-optimal strategy in both cases has one switching point. The gain in time compared to traditional strategy in this case is large enough and it is equal to 285 times for the gradient method and 200 times for the DFP method.

Table IV. Results of the analysis of quasi-optimal design strategies for the three-stage transistor amplifier with a variable integration step

N	Method	Optimal control vector U(u1,u2,u3,u4,u5,u6,u7)	Iterations number	Switching points	Time (sec)	Gain in time
1	Gradient	(1111111); (1111101)	363	350	1.127	285
2	DFP	(1111111); (1110111)	69	66	0.322	200

The results of the design of nonlinear passive circuits show that TSD is not the best in time. Quasi-optimal strategy of design provides a significant gain in design time compared to the TSD and this gain increases with the number of nodes in the network. This conclusion is illustrated in Figure 6 in the form of dependencies of time gain on the parameter M for a variable optimal integration step the main system.

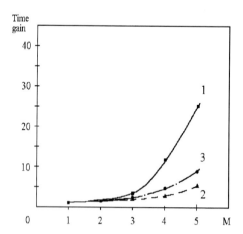

Figure 6. Time gain as a function of the number of node scheme: 1-gradient method, 2 - Newton's method, 3 - a method of DFP.

The results of the design of nonlinear active circuits show that TSD is not the best in computer time. Quasi-optimal strategy of design provides a significant gain in time design compared to the TSD and this gain increases with increasing size and complexity of the circuit. This conclusion is illustrated in Figure 7 in the form of graphs of the gain in time for designing the amplifiers with different number of transistor's cascades for the two optimization methods, the gradient method and the DFP.

The results for the examples that have been analyzed show that the potential gain in time, which was obtained by means of a quasi-optimal strategy of design, increases with the size and complexity of electronic circuits. These potential opportunities exist due to almost infinite number of different design strategies that appear in a generalized design methodology. The number of different strategies within the structural basis of the process of design increases exponentially with the number of nodes in the circuit. In the latter example the structural basis consists of 128 different strategies. Already these strategies give the gain in computer time tenfold. At the same time, additional optimization, by changing the control vector indicates that there are strategies that we called a quasi-optimal that can provide the gain in computer time hundreds or thousands.

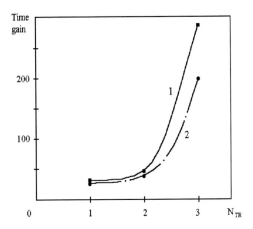

Figure 7. Time gain as a function of the number of cascades of transistor amplifier: 1 - gradient method, 2 - a method of DFP.

Effect of Acceleration of the Process of Designing

The basic concept of the new approach is the introduction of control functions, which generalize the design process and allow you to control this process to achieve the optimum point of objective function of design in the fastest possible time. As shown in [23-24], the potential gain of computer time, which can be obtained for the optimal design strategy, compared with the traditional one increases with the size and complexity of electronic circuits. However, the potential gains can be realized in practice only in the case of constructing an algorithm allowing allocating the optimal trajectory of designing process.

On the basis of developed methodology in previous sections, this section analyzes the special effects arising in the process of circuit designing based on a generalized methodology. The study is carried out starting with the simplest non-linear circuit with one node and two parameters ($N = 2$), which has no practical significance, but serves as a good model for understanding the processes occurring in the designing of circuits based on the developed methodology. Then we analyze the N-dimensional problem. All considered examples demonstrate that there is a certain effect, which may be called the effect of accelerating the designing process, and this effect was appeared resulting from the different behavior of the trajectory design, with different control functions.

In addition to the concept a strategy of design, introduced in earlier sections, we introduce the concept a trajectory of design corresponding to a given strategy. Under the trajectory of designing we understand the dependence of the vector of state variables of the X as a function of current designing time, serving as a parameter. Since the vector X belongs to the N-dimensional space ($X \in R^N$), the trajectory design is also in space R^N. Graphically, the trajectory design represented by a line in space R^N. Graphically, the trajectory of designing represented by a line in space R^N. The starting point of this line X^0 corresponds to the initial approximation to the solution of designing problems and is determined by the designer. The end point of the trajectory is a solution to the designing problem, since at this point we have all of the independent parameters of the electronic system that is, achieving the goal of designing. In the case where the vector X depends only on two variables, one independent variable and another dependent one, the trajectory of designing is a curve on the plane and can easily be reproduced graphically. In all other cases, when $N > 2$, is useful to depict the projections of the trajectory design in various planes. Most informative were the projections on those planes for which a one variable belongs to the set of independent variables X' and the other - the set of dependent variables X''.

Let us study the effect of acceleration for two-dimensional problem. Recall that the vector of parameters of mathematical models of the chain X can be divided into two parts: $X = (X', X'')$ where $X' \in R^K$ is the vector of independent variables, K is the number of independent variables, $X'' \in R^M$ is the vector of dependent variables, M is the number of dependent variables $N = K + M$.

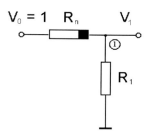

Figure 8. Simplest nonlinear circuit with one node.

Simplest non-linear electronic circuit with one node is shown in Figure 8. We believe that the resistance R_n is defined by the nonlinear dependence in the form: $R_n = r_{10} + b_n \cdot (V_0 - V_1)^2$. For this example, we define only two parameters: the resistance R_1 as an independent parameter ($K = 1$) and the nodal voltage V_1 as a dependent parameter ($M = 1$). It is also assumed that all the resistance or conductivity are positive. To automatically perform the latter requirement can be used as it was previously shown the following definition of the vector X: $X = (x_1, x_2)$, where $x_1^2 \equiv R_1$, $x_2 \equiv V_1$. In this case the equation corresponding to the circuit, given by the following formula:

$$x_2 = x_1^2 / \left(x_1^2 + r_{10} + b_n \left(V_0 - x_2 \right)^2 \right) \tag{30}$$

The model of circuit in standard form can be written by the following equation:

$$g_1(X) \equiv \left(x_1^2 + r_{10} + b_n \left(V_0 - x_2 \right)^2 \right) x_2 - x_1^2 = 0 \tag{31}$$

By its physical meaning the circuit is a voltage divider. The task of designing this circuit is to find the value of the independent variable x_1 (R_1), enables one to obtain on the output correct voltage V_{out}. The objective function of the design process we define as: $C(X) = \left(x_2 - k_V \right)^2$. In this case, the vector control function contains only a component u_1, since there is only one dependent parameter x_2. Trajectories of designing in this case are curves in two dimensions, using the numerical solution algorithm. At the same time, the numerical analysis of this scheme makes little sense for design purposes, since there is exist an analytic solution of this problem, such as the method of Lagrange multipliers in the case of a fixed value of the function u_1. However, we have carried out numerical solution of this problem in order to identify the essential properties of new effect that accelerate the designing process. The main features of this effect are seen also in all other examples.

Optimization procedure and the model circuit in accordance with the proposed methodology are given by the following equations:

$$x_i^{s+1} = x_i^s + t_s \cdot f_i(X,U), \qquad i = 1,2,...,N \tag{32}$$

$$\left(1 - u_j \right) g_j(X) = 0, \qquad j = 1,2,...,M \tag{33}$$

where $N = 2$, $M = 1$, U is the control vector, consisting in this case, one component, and functions $f_i(X,U)$ for $i = 1,2$, representing the components of directional movement, determined by the chosen method of optimization. These functions, for the gradient method, can be defining the following formulas:

$$f_1(X,U) = -\frac{\delta}{\delta x_1} F(X,U) \tag{34}$$

$$f_2(X,U) = -u_1 \frac{\delta}{\delta x_2} F(X,U) + \frac{(1-u_1)}{t_s} \left[-x_2^s + \eta_2(X) \right] \tag{35}$$

where $F(X,U)$ is the generalized objective function, defined in this case by the formula $F(X,U) = C(X) + \frac{1}{\varepsilon} u_1 g_1^2(X)$, $\eta_2(X)$ is an implicit form of the function that defines

192   A.M. Zemliak

the system of equations (33), $\left(x_2^{s+1} = \eta_2(X)\right)$ and the operator $\delta / \delta x_i$ for $i = 1,2$ is

defined as follows: $\dfrac{\delta}{\delta x_1} F = \dfrac{\partial F}{\partial x_1} + \dfrac{\partial F}{\partial x_2} \dfrac{\partial x_2}{\partial x_1}$, $\dfrac{\delta}{\delta x_2} F = \dfrac{\partial F}{\partial x_2}$.

As shown in [22] the system (32) can be rewritten in a continuous form as a system of differential equations:

$$\frac{dx_i}{dt} = f_i(X,U) \qquad (36)$$

The basis of different strategies of design, which is defined by means of the control's vector U, a constant throughout the design process consists in this case of two strategies, with $u_1 = 0$ and $u_1 = 1$. The first strategy is the traditional strategy of design (TSD) and the second - a modified traditional strategy of design (MTSD).

The results of the designing for the different parameter of nonlinearity for the three optimization methods are summarized in a comparative Table V.

Table V. Results of the analysis of the full structural basis of different strategies of design for the three optimization methods

N	Functions of control	Gradient method		Newton method		DFP method	
		Iterations number	Time (sec)	Iterations number	Time (sec)	Iterations number	Time (sec)
1	0	9	0.001018	7	0.001653	7	0.001427
2	1	72	0.006768	11	0.002931	12	0.002406

The results of the designing when the parameter of nonlinearity $b_n = 1,0$ for the three optimization methods, the gradient method, Newton's method and the method of the DFP are summarized in a comparative Table V for the TSD ($u_1 = 0$) and MTSD ($u_1 = 1$) and an initial guess for the vector X: $X^0 = (1,1)$. TSD is preferred for this example, since it allows getting a solution faster than the MTSD for all optimization methods. A slight decrease of time of design (up to 5%) can be obtained for example for the gradient method when the control function u_1 is changed in the process of movement from the value 0 to the value 1 in the step 7. Trajectory for the two design strategies that correspond to the gradient method of optimization of Table V and the initial approximation $X^0 = (1,1)$ for three values of the nonlinearity parameter b_n, 10^{-5}, 1.0 and 5.0, are shown in Figure 9 (a), (b), (c).

Solid lines correspond to the TSD, dashed lines correspond to MTSD. As is evident from the graphs, both strategies lead to the same values for the variables x_1 and x_2 at the end of the design process, which is an important prerequisite for comparing different trajectories. In addition, the trajectories for both strategies lie in physically real space, as all of the values of both variables x_1 and x_2 are positive along the trajectories.

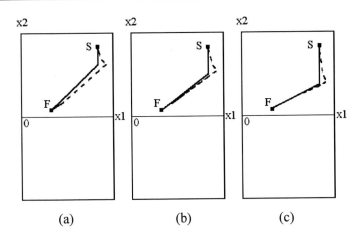

Figure 9. Trajectories for TSD (solid) and MTSD (dashed) with $x_2 = 1$ for three values of nonlinearity b_n: a) 10^{-5} b) 1.0 and c) 5.0.

Different behavior of the trajectories observed when selecting a negative value of the initial approximation for a variable x_2, such as -1, then there is an initial approximation $X^0 = (1,-1)$. Within the meaning of this means that the initial value of the output voltage of divider is taken negative, although the input voltage is assumed to equal 1. Trajectories corresponding to this case are shown in Figure 10 (a), (b), (c) for three values of the nonlinearity b_n, 10^{-5}, 1.0 and 5.0.

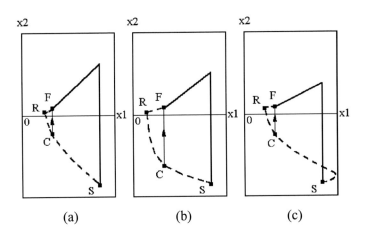

Figure 10. Trajectories for TSD (solid) and MTSD (dashed) with $x_2 = -1$ for three values of nonlinearity b_n: a) 10^{-5}, b) 1.0 and c) 5.0

Trajectories corresponding to TSD remain virtually unchanged. In the first case, with positive coordinate x_2, (Figure 9) exhibits a jump downward trajectory from the point of initial approximation to the line corresponding to an adjusted solution of (33), and in the second case, there is a jump on the same line from the bottom up. Since the jump is performed instantaneously, then the time in both cases equally. The situation is different with the trajectory of MTSD. For a negative initial value of variable x_2 ($x_2 = -1$), the first part of the trajectory lies in a physically unreal subspace with negative value of variable x_2 but the second part in the positive subspace. The number of iterations and CPU time, for example for the gradient method, is equal 288 and 0.027 seconds. It should be noted that the motion of the current point in the first part of the trajectory from the initial point S till point R is carried out fairly quickly and then slows down, as we approach the end point. The total time for MTSD is increased by 4 times in this case with respect to previous case with a positive initial value of variable x_2. This means that such a change in the starting point of the design affects the characteristics of the design process if the control's vector remains constant throughout the trajectory. It is also important to note that the trajectories of two different strategies suited to the end point of the design process F from opposite sides. This creates a unique opportunity to accelerate the process of switching control function u_1 with value 1 to the value 0 at some point C, is the projection of the end point F on the trajectory corresponding to the MTSD. In this case, the optimal strategy and the corresponding optimal trajectory have two parts. The first part is the curve SC, corresponds to $u_1 = 1$ and MTDS and lies in a physically unreal space, because the output voltage is negative, although the input voltage is equal to 1. At point C the value of the control function is changed to 0 and thus made the jump or directly to the endpoint of F, or pretty close to it, that depends on the step of integrating the system (36) and a specified accuracy. The second part of the trajectory starting at point C and the corresponding value of $u_1 = 0$ and TSD thus degenerates to a jump with only one step or even a few additional steps, corresponding to the TSD. In this case, there is a significant acceleration effect of the design process. This effect is observed at all values of the nonlinearity parameter b_n. In the first case, where $b_n = 10^{-5}$, electronic circuit is essentially linear, in the third case, the nonlinearity parameter is large and the trajectory corresponding to MTSD, more nonlinear, but regardless of this, the acceleration effect is observed for all values of nonlinearity.

Data that correspond to the nonlinearity parameter $b_n = 1.0$, the initial approximation $X^0 = (1,-1)$ and the three different optimization methods are presented in Table VI for the quasi-optimal design strategy.

Table VI. Results of the analysis of quasi-optimal strategies of design for passive nonlinear circuits with a single node in the presence of the acceleration effect

N	Method	Optimal control vector	Iterations number	Switching point	Time (sec)
1	Gradient	(1 0)	2	1	0,0002071
2	Newton	(1 0)	2	1	0,0005025
3	DFP	(1 0)	2	1	0,0004043

This quasi-optimal strategy consists of two parts for all optimization methods, both the first part and second part consist of only a one step. Quasi-optimal strategy starts with the MTSD for which $u_1 = 1$ and after the first step the function u_1 switches to the value 0, then there is a corresponding TSD. The resulting gain in time for quasi-optimal strategy compared with the TSD by means of an acceleration effect achieves 4.91 times for the gradient method, 3.29 times for Newton's method, and 3.53 times for the method of the DFP.

The effect of acceleration was investigated in [33, 34] and should be noted that he was observed for the more complicated examples too. However, in this case, the trajectory of the process lie in the N-dimensional space, and it is necessary to analyze the different projections of N-dimensional curves.

Let us define N-dimensional problem. Consider an electronic circuit with three nodes, shown in Fig. 11.

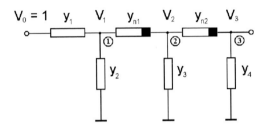

Figure 11. The scheme of nonlinear passive electronic circuit with three nodes.

Vector X is defined by seven components, the following formula: $x_1^2 = y_1$, $x_2^2 = y_2$, $x_3^2 = y_3$, $x_4^2 = y_4$, $x_5 = V_1$, $x_6 = V_2$, $x_7 = V_3$. The structural basis of different strategies of design consists of 8 different strategies. Analysis of various strategies of design for the structural basis conducted in [33] for both a constant integration step and variable optimal integration step of the basic system (21).

With a constant integration step, using a gradient optimization method, the following results have been obtained: in the case of TSD (U = (000)) the number of iterations is equal to 4073 and CPU time is 5.269 seconds, with the initial vector $X^0 = (1,1,1,1,1,1,1)$. Optimal, or rather quasi-optimal strategy in this example has two switching points of the vector of control functions U, on step 10 and on step 2000. Vector of functions of control have the following sequence (101) (110) (111), and the total number of iterations equal to 2009. CPU time was 0.9084 seconds, and a time gain, it is easy to calculate, 5.77 times.

Using the method of optimizing the DFP, yielded the following results: in the case of TDS number of iterations equal to 284 and the CPU time of 0.673 seconds, if the initial vector $X^0 = (1,1,1,1,1,1,1)$. Quasi-optimal strategy for this example would have two switching points of the vector of control functions U, in step 5 and in step 59. Vector control functions have the following sequence (101) (111) (110), and the total number of iterations equal to 173. CPU time was 0.1463 seconds, and the gain in time 4.6 times.

Investigation of possibilities of the effect of acceleration for the analyzed circuit showed that the effect occurs when specifying different initial values of the vector of state variables X, if there is at least one negative value among the components of vector X^0, the corresponding dependent variables. This effect is also presented at different combinations of strategies, which were chosen from the fundamental basis.

It is interesting to note that all trajectories corresponding to the strategies of structural basis can be divided into two parts. Figure 12 shows the projections of the trajectories of the basic strategies at the plane $x_4 - x_7$ corresponding to the initial vector $X^0 = (1,1,1,1,1,1,-1)$.

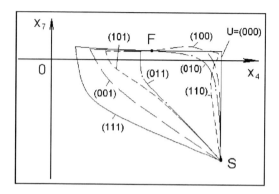

Figure 12. Projections of trajectories in phase space for the circuit with three nodes.

Complete basis consists of eight strategies. We can divide the set of these strategies into two subsets. One of these subsets includes trajectory TSD, $U = (000)$ and three other trajectories like TSD, namely, the strategy with a vector of control (010) (100) and (110). The second subset includes the trajectory of MTSD, $U = (111)$ and three other similar MTSD with the vector of control (001) (011) and (101). The main difference between these two groups lies in the fact that their representatives approaching to the final point F from opposite directions, which ultimately creates an opportunity for the emergence of the acceleration effect. All the trajectories of the second group may be selected as candidates for the construction of the first part of the quasi-optimal trajectory, and vice versa, the trajectories of the first group are suitable for the jump to the end point or close to it, which is a characteristic feature of the acceleration effect. One of the possible options for obtaining quasi-optimal strategy of design was implemented based on two strategies: TSD with control vector (000) and MTSD with control vector (111). Table VII shows the results of this implementation, with a constant step of integrating the system (21) and two methods of optimization, the gradient method and the DFP.

Table VII. Quasi-optimal strategy of designing for nonlinear passive circuit with three nodes with effect of acceleration

N	Method	Optimal control vector U (u1, u2, u3)	Iterations number	Switching points	Time (sec)	Gain in time
1	Gradient	(111); (000); (111)	1598	1251; 1252	0.0128	17.6
2	DFP	(111); (000); (111)	87	40; 41	0.00224	13.2

The initial value of the vector of variables X is chosen as follows: $X^0 = (1,1,1,1,1,1,-1)$. Quasi-optimal strategy of designing has two switching points of the vector of control functions, from (111) to (000) and from (000) to (111) for both optimization methods. In the case of the gradient method of optimization, these switching carried out on the steps of integrating the 1251 and 1252, but in the case of the method of DFP at the steps number 40 and 41. The gain in time compared to the TSD was 17.6 times for the gradient method and 13.2 times for the method of the DFP. This corresponds to an additional acceleration of the design process to 3 times for the gradient optimization method, and 2.87 times for the method of DFP compared with those previously obtained.

The three-stage transistor amplifier. Figure 5 shows a scheme of a three-stage transistor amplifier, which was analyzed earlier without the possible effect of acceleration. Gain in time for the quasi-optimal strategy amounted to 285 times using a gradient optimization method, and 200 times using the method of the DFP.

Acceleration effect allowed us get a more significant gain in time. In this case the structure of the vector of control is determined by the following values of components: (1111111), (0000000), (1111111) for the gradient optimization method, and for the method of the DFP. Table VIII shows the results obtained for the quasi-optimal strategies for the two switching points of control vector in the presence of the acceleration effect with a variable step integration and two methods of optimization: the gradient method and the DFP.

Table VIII. Quasi-optimal strategies of designing for three-stage transistor amplifier in the presence of the acceleration effect with a variable integration step

N	Method	Optimal control vector $U(u1,u2,u3,u4,u5,u6,u7)$	Iterations number	Switching points	Time (sec)	Gain in time
1	Gradient	(1111111); (0000000); (1111111)	247	10, 11	0.518	620
2	DFP	(1111111); (0000000); (1111111)	29	2, 3	0.135	477

Gain in time for the quasi-optimal strategy of designing in the presence of an additional acceleration effect as compared with TSD is illustrated in Figure 13 for the two optimization methods in the form of dependencies of the gain in time versus the number of cascades N_{TR} of amplifier. For the gradient method, when $N_{TR} = 3$, the gain in time is equal 620, while for the method of the DFP, this gain is equal to 477. Additional gain in time due to the effect of acceleration was 2.17 for the gradient method, and 2.38 for the method of the DFP.

Analysis of all these examples gives us possibility to make a definite conclusion that the effect of acceleration for the designing process is the first important element of quasi-optimal in time algorithm.

Studies have shown that the structure of the quasi-optimal trajectory must include conditions to achieve the additional effect of acceleration the designing process. One of these conditions is the correct choice of initial approximation, i.e. the presence into the initial value of the vector of variables X at least one negative component among the originally dependent variables, in our case, the nodal voltages.

The second condition is the correct choice of sequence of the strategies of designing in the process of constructing the optimal trajectory. We have in mind the choice in the initial phase of one of the strategies that has a trajectory similar to MTSD, and then the changing it to a strategy that has a trajectory similar to TSD. Additional changes of the vector of control can approach closer to the optimal strategy of designing with respect to time.

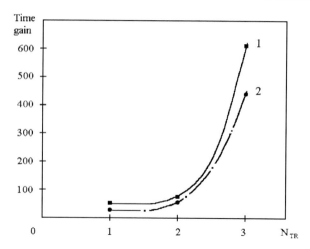

Figure 13. Gain in time as a function of the number of stages, taking into account the acceleration effect. 1 - gradient method, 2 - method of DFP.

The third condition is the ability to determine the exact moment of switching of control functions. This problem can be solved with the involvement of approximate methods of the theory of optimal control.

Sufficient conditions to obtaining the acceleration effect. In the previous section it is shown, that under certain conditions there is a possibility to realize effect of acceleration of process of designing of electronic circuit. It is shown, that this effect appears if among components of a vector of initial approximation X^0 to choose negative one or several components concerning to dependent variables. As dependent variables in physical sense are nodal voltages it means, that in an initial point of process of designing it is necessary to choose negative some nodal voltages to obtain the effect of acceleration.

Since the effect of acceleration is observed for all the examples, these conditions can be formulated as sufficient conditions for obtaining the effect of acceleration of the designing process. The question is whether these conditions are necessary, subject to further investigation. The acceleration effect of the designing process can be considered the first practical element of constructing an optimal in time algorithm. The problem of selecting the starting point of the designing process is one of the important problems in constructing real optimal or quasi-optimal algorithm. In the new generalized formulation of the designing process the task correct choice of initial approximation is even more significant.

Analysis of the effect of acceleration of the designing process for a simple nonlinear circuit with one node has been given above. Let us consider in more detail the processes occurring in the choice of different initial approximations. In this case, the vector of state variables X has two components $X = (x_1, x_2)$, where $x_1^2 = R_1$, $x_2 = V_1$ and the design process is carried out in the space of two dimensions. Structural basis for this example consists of two strategies: TSD and MTSD. Changing the starting point of the designing process has virtually no effect on the behavior of the trajectories of the TSD, since in this case is always carried out the jump from the initial point on the line, which is a correct solution of (33). In contrast to the TSD, strategy MTSD undergoes a significant dependence on the initial approximation. As shown previously, to obtain the acceleration effect, we must select the start point of the designing process, which has negative coordinate x_2. Figure 14 shows a family of trajectories (a phase portrait), corresponding to the designing of the circuit shown in Figure 8, by applying the MTSD ($u_1=1$), for some initial negative values of coordinate x_2.

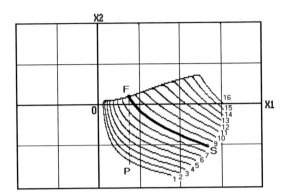

Figure 14. The family of trajectories MTSD with different initial values of coordinate x_2.

The curves start at different initial points, but all of them finish at the same end point of the designing process F, where the minimum objective function $C(X)$ is achieved. The starting points were chosen on the arc of a circle. Special curve S-F, that is marked by a bold line can be named as separatrix [35, 36] that separates the trajectories that are the candidate for the realization of the acceleration effect (curves lying below the separatrix) and curves that cannot produce the acceleration effect (curves lying above the separatrix). It is evident that in spite of the negative value of the coordinates x_2 the trajectories from 9 to 15 cannot produce the acceleration effect.

It is obvious that the projection of the final point F on the curve of the first group defines the switching point of the control function u_1 necessary to achieve the acceleration effect. All the curves of the first group (1-7) are come nearer to the final point F on the left, while the trajectories of the second group (9-16) reach the point F on the right. Relative computing time τ for the trajectories of Figure 14 is shown in Figure 15 as a function of curve number n.

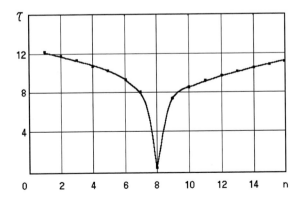

Figure 15. Relative computing time for different trajectories MTSD.

Trajectory, which coincides with the separatrix S-F, has a minimal computer time because we are "guessing" in this case, the coordinates of the starting point of the designing process and the current point moves to the final point at the quickest possible way. At the same time, this trajectory cannot be the basis for the acceleration effect and for the obtaining an optimal algorithm of the designing, because the projection of the final point F on this trajectory is the itself point of F, but, as is well known, just near the final point, the optimization process significantly slows. At the same time, the current point of the designing process moves quickly, being far from the final point F. It is clear that the computer time for the strategies that correspond to the initial points 7 and 9 is more than for the point S, corresponding to the start point of the separatrix, but at the same time, less than for strategies with other starting points. Computing time increases with the distance increasing between point S and any other starting point. Relative computing time that corresponds to the trajectories starting at the initial point S and in point 2 for example is differed by 12 times.

In Figure 14 is shown that only trajectories that lie below the separatrix are the basis for the first part of the optimal trajectory. The following jump from a point that is a projection of the endpoint of F is the second part of the optimal trajectory. This jump is realized by changing the values of the control function u_1 from 1 to 0 at the right moment and at precisely the moment when the current point moving along the trajectory reaches the point of intersection of the trajectory with the vertical line F-P. Relative computing time τ corresponding to the trajectory, which realize the acceleration effect (on the basis of curves 1-7, Figure 14), is shown in Figure 16 as a function of curve number n.

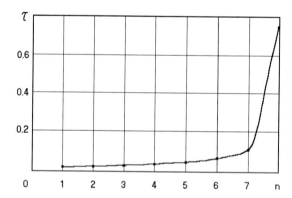

Figure 16. Relative computing time τ for the optimal trajectories with different initial conditions.

It should be noted that the maximum point of the curve corresponding to the number 8, is at the same time, the minimum point of the curve in Figure 15. There is a significant difference in computer time for strategies that start at S and, for example, at 1. For these points the difference in time of more than 35 times. Can show that the trajectories from 9 to 16 can also be optimized with respect to time, but in this case we can reduce the CPU time only on 10-15%. Figure 16 shows that the computing time increases for curves 1-7 when the starting point is close to the separatrix, and on the contrary, one can observe a greater acceleration if the initial point moves off away from the separatrix. This effect is associated with different length of the trajectories from their initial point till a projection along the line FP, obtained by the choice of initial points on the arc of a circle, as well as already noted due to slowdown of the optimization process when approaching to the endpoint. Consequently, the choice of the initial point of the design process with a negative value of the coordinate and the value of this coordinate, corresponding to the position below the separatrix, are the sufficient conditions for obtaining the acceleration effect.

Separatrix of the first and second kind. Necessary and sufficient conditions for obtaining the acceleration effect. A more detailed analysis shows that the condition for choosing the initial point with negative coordinates x_2 is not really necessary. Figure 17 shows the phase portrait of the designing process, corresponding to MTSD for the same circuit in Figure 8, but for all possible values of the coordinate x_2. It should be emphasized that the trajectories corresponding to the MTSD, constitute the first part of a possible quasi-optimal trajectory planning. Therefore, their behavior and determines the presence or absence of opportunities for the obtaining of an acceleration effect.

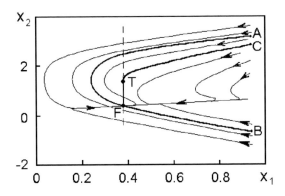

Figure 17. Phase portrait for the trajectories of MTSD in the extended area and the structure of the separatrices.

Phase portrait of the designing process on the basis of MTSD includes two types of special objects, which we called separatrices. These lines are marked in Figure 17 by thickening. First, we note the separatrix AFB, which separates the trajectories approaching to the end point from the left side and from the right side. This seperatrix called as a separatrix of the first type. Separatrix of the second type CTFB divides the phase space into two different subspaces. Points of subspace that lies inside the separatrix do not lead to trajectories that allow realizing the acceleration effect. Conversely, points lying outside the separatrix correspond to trajectories that can produce the acceleration effect, because in this case it is possible to jump directly to the end point F of the designing process, or close to it. In this case, the jump till the end point F can be made both from below and from above the separatrix CTFB on the continue of the line TF. These geometric conditions are the necessary and sufficient conditions for the existence of the acceleration effect.

Such a clear explanation is possible only in two-dimensional case. Cases N-dimensional measurements are analyzed below. The separatrix become a line or surface of a three-dimensional space when $N = 3$, and we can study their projections on the plane only. In the case $N > 3$ the separatrix become a line of N-dimensional space, surface or hypersurface. Structure of the separatrix is very complex, and the number of possible two-dimensional projections drastically increases. Clearly, in this case it is necessary to confine oneself of the smallest number of the most informative projections. The question is what type of the projections are the most informative for the designing process. The analysis of various examples has shown that two-dimensional projections are more important on the plane conductivity-voltage.

Results of designing one and three cascaded amplifiers in the form of projections of phase portraits on the corresponding planes are shown in Figure 18 for a one-stage amplifier, and in Figure 19 for a three-stage amplifier.

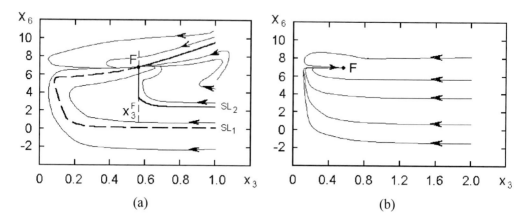

Figure 18. Phase portraits and the structure of the separatrices for one-stage transistor amplifier: (a) $x_i^0 = 1.0$, $i=1,2,\ldots,K$ ($K=3$), (b) $x_i^0 = 2.0$, $i=1,2,\ldots,K$ ($K=3$).

In the case of one-stage amplifier has three independent variables y_1, y_2, y_3 ($K=3$) and three dependent V_1, V_2, V_3 ($M=3$) when using the TSP. The vector X includes 6 components: $x_1^2 = y_1$, $x_2^2 = y_2$, $x_3^2 = y_3$, $x_4 = V_1$, $x_5 = V_2$, $x_6 = V_3$. One of the most informative projections are the projections on the plane $x_3 - x_6$. Figure 18 shows the projections of some of the trajectories of the designing process, and the separatrix of the first and second types.

First figure shows the behavior of projections of trajectories on the plane $x_3 - x_6$ with the initial values of independent variables that equal: $x_i^0 = 1,0$, $i=1,2,3$, and the second with the initial values of independent variables that equal: $x_i^0 = 2,0$, $i=1,2,3$. First, it is possible to ascertain a significant difference in the behavior of phase trajectories for passive and active circuits. Secondly, the form of phase trajectories and separatrix depends strongly on the initial values of independent variables. Separatrix of the first and second kind SL_1 and SL_2 are clearly defined and the behavior of their projections on the plane $x_3 - x_6$, in the case $x_i^0 = 1,0$, clearly explains the presence or absence of the acceleration effect. In contrast to this case, the projection of separatrices disappear into the plane $x_3 - x_6$ with the initial values $x_i^0 = 2,0$. This means that the acceleration effect in the first case, when $x_i^0 = 1.0$, can be carried out in the region corresponding to the projections of points, located on the outside of the separatrix of type 2. In contrast, in the second case, when the initial values of the independent variables are equal to 2.0, the acceleration effect is realized for all initial values of coordinates, because all trajectories include the ability to "jump" into the end point of the designing process. This means that the separatrix in this case disappears.

For the three-stage transistor amplifier, vector X has 14 components, seven independent components and seven dependent components for the TSD. Figure 19 shows two-dimensional projections of phase portraits of the designing process by MTSD for two cases: (a) when the initial values of independent variables $x_i^0 = 1.0$, $i = 1, 2, ..., K$ ($K = 7$) and (b) when the initial values of independent variables $x_i^0 = 2.0$, $i = 1, 2, ..., K$ ($K = 7$).

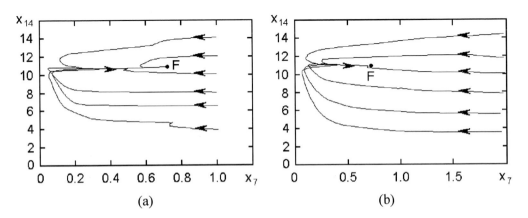

Figure 19. Phase portraits and the structure of the separatrices for three-stage transistor amplifier: (a) $x_i^0 = 1.0$, $i = 1, 2, ..., K$ ($K=7$), (b) $x_i^0 = 2.0$, $i = 1, 2, ..., K$ ($K=7$).

As seen from the figures, the separatrix in this case does not appear, that is, the acceleration effect may exist in all space. This means that a further increase in circuit complexity can reduce and eventually give to the disappearance of the space within which there is no condition for the effect of acceleration.

The optimal choice of the starting point of the design process allows get us the acceleration effect with high probability. Analysis of the trajectories for different strategies of designing shows that the concept of separatrices is useful for understanding and identifying the necessary and sufficient conditions for the existence of the effect of accelerating the design process. Separatrix divides the phase space projection onto the area in which it is possible to obtain the acceleration effect, and the area where this effect does not exist. The first area can be used to construct the optimal trajectory of designing. Selecting the starting point of the design process from outside the region bounded by the separatrix is a necessary and sufficient condition for the existence of the effect of acceleration. Separatrices are as a hypersurface and has a complicated structure in general. However, the real situation is simplified in the most important case of active nonlinear circuits because of narrowing of the field inside the separatrix or the complete disappearance due to sufficiently large initial values of the independent variables. This means that the acceleration effect can be realized almost always for enough for complex circuits. This effect allows reduce additionally of the CPU time and may serve as one of the basic ideas for constructing an optimal or quasi-optimal algorithm of designing.

The Lyapunov function of the design process

To obtain the optimal sequence of switching points in the design process we need to develop a specific criteria, which depends on the intrinsic properties of design strategy. The task of finding an optimal design strategy with respect to time related to the more general problem of convergence and stability of each trajectory of designing. Based on experience we can say that the time of designing for each strategy depends of the properties of convergence and stability of the trajectory corresponding to this strategy. One of the most common approaches to stability analysis of dynamic systems is based on direct method of Lyapunov [37, 38]. We have defined the procedure for design of electronic circuits as a controllable dynamic system [39, 40]. The design process is considered as bringing the system in steady state. In this case, at steady state, minimizes the objective function $C(X)$ that is achieved all design objectives. The result of the optimal design is to minimize the objective function $C(X)$ for the minimum possible CPU time. In this case, the main goal of optimal control can be defined as the problem of minimizing the transition time of the process of bringing the system in steady state. Since bringing the system in steady state, i.e. the implementation of the process of designing electronic circuits, possibly through various strategies, we must find a strategy and the corresponding trajectory realizing the transition process in the fastest possible time. It is important to analyze the stability of each trajectory and characteristics of the transition process, which is the design process. This analysis can be done based on the Lyapunov direct method. It is proposed to introduce the concept of Lyapunov function of the design process and use it to study the structure and properties of the optimal algorithm design and, in particular, to find the optimal position of the switching points the control vector U.

There is some freedom in choosing the Lyapunov function due to non-uniqueness of its shape. Let's define a Lyapunov function of the design process by the following formula:

$$V(X) = \sum_i (x_i - a_i)^2 , \tag{37}$$

where a_i is a steady, final value of the coordinate x_i. According to its meaning, the set of all coefficients a_i is the main result of the design process, since the objective function $C(X)$ has a minimum for these values of the coefficients a_i, i.e. all design objectives are achieved. Clearly, these coefficients are determined only at the end of designing. Instead, the variables x_i we can may determine the other variables $y_i = x_i - a_i$. In this case formula (5.6) takes other form:

$$V(Y) = \sum_i y_i^2 \tag{38}$$

where the components y_i compose the vector Y.

The design process can be written to the new variables y_i in the same form. Function (38) meets all the conditions of a standard definition of Lyapunov function. Indeed, this function is piecewise continuous and has piecewise continuous first partial derivatives. In addition there are three basic properties of functions: 1) $V(Y) > 0$, 2) $V(0) = 0$, and 3) $V(Y) \to \infty$, when $\|Y\| \to \infty$. In this case it is possible to analyze the stability of equilibrium, i.e. point $Y = 0$ on the basis of Lyapunov's theorem. On the other hand, the stability of the point $A = (a_1, a_2, ..., a_N)$ can be analyzed on the basis of the definition (37). It is clear that both problems are identical. The disadvantage of (37) lies in the fact that the point $A = (a_1, a_2, ..., a_N)$ is unknown, since it can only be obtained at the end of the design process. That is, we can analyze the stability of various design strategies on the basis of (37) in case when we have the solution of the problem, i. e., the point the A, is already found by other way. On the other hand, it is interesting to control the stability of the process during the procedure of optimization. In this case, we want to define another form of Lyapunov function, which would not depend on a finite fixed point A in explicit form. Stationary point corresponds to 0 in coordinates y_i and corresponds to A in coordinates x_i. Let's define a Lyapunov function by the following formulas:

$$V(X,U) = [F(X,U)]^r, \tag{39}$$

$$V(X,U) = \sum_i \left(\frac{\partial F(X,U)}{\partial x_i} \right)^2, \tag{40}$$

where $F(X, U)$ is a generalized cost function of the designing process and the degree $r > 0$. Both formulas, under certain conditions, define a Lyapunov function, which has properties similar to the function (38) in a sufficiently large neighborhood of the stationary point and, thus, in both formulas, a dependence on the vector of control U is observed. Indeed, in relation to the function (39) we can say that in a stationary point, i. e. at the end of the design process, the value of this function is zero if the objective function of the designing process $C(X)$ at this point is zero. Function (39) is positive definite at all points other than stationary one, because it is assumed that the function $C(X)$ is not negative definite. The function $V(X,U)$ increases indefinitely when the point X moves off the stationary point. Formula (40) can also define the Lyapunov function, if the derivatives $\partial F / \partial x_i$ are zero at the stationary point $A = (a_1, a_2, ..., a_N)$, so that $V(A,U) = 0$. On the other hand, $V(X, U) > 0$ for all X and, finally, Lyapunov function, which defined by (40), there is also a function of the vector U, since all the coordinates x_i are functions of the vector U. The property 3) of the Lyapunov function is not proved only, since the unknown behavior of the function $V(X, U)$ at $\|X\| \to \infty$. However, it can be assumed on the basis of practical experience that the function $V(X, U)$ increases in a fairly large neighborhood of the stationary point.

In accordance with the method of Lyapunov, information about the stability of the trajectory of the system is connected with the time derivative of Lyapunov function. Direct calculation of the time derivative of Lyapunov function $\dot{V} = dV/dt$ gives a possibility to define the stability of a dynamical system. Design process and the corresponding trajectory are stable, if this derivative is negative. On the other hand, the direct method of Lyapunov, is known to give sufficient conditions for stability, but not necessary. This means that the process is stable or becomes unstable in the case of a positive derivative. If at some points of the trajectory of designing the derivative $\dot{V}$ becomes positive, it does not mean the appearance of instability of the trajectory design at these points. Only in the case of a positive derivative $\dot{V}$ on a set of positive measure of sufficient size can predict the appearance of instability that is manifested in the increase, although perhaps not large, the Lyapunov function, not its reduction. In this case we can decide that, from this moment, the objective function of the design process does not decrease but increases. If such behavior occurs away from the stationary point, then it means that the design process does not converge, i. e. a solution cannot be obtained on this trajectory. In this case, we need to change or starting point of the designing process or the strategy itself or both at once. In the case of a positive derivative $\dot{V}$ at the end of the design process, not far from the stationary point, we can speak of a significant slowdown in the design process. This strategy begins "mark time" and cannot provide the required accuracy of the solution, resulting in a significant increase in designing time. This effect is well known in practical designing. If in this case, the accuracy is unacceptable, we need once again change the strategy of designing or at least a starting point. In the traditional designing we need change the starting point, or perhaps the optimization method. Detailed analysis of the behavior of the Lyapunov function and its derivative for different strategies of designing allows you to select promising strategies and discard wittingly unsuccessful. This analysis also reveals, at least on a qualitative level, the relationship between design times and, as we have determined the main parameters of the design process, namely the Lyapunov function and its time derivative.

Time's derivative of Lyapunov function and its application. In accordance with the theory of Lyapunov's direct method, the information on the stability of trajectory and, in our case, about the CPU time is correlated with the time derivative of Lyapunov function. In the previous section shows that the Lyapunov function of the design process together with its derivative could provide a helpful source for finding promising in terms of the minimum CPU time, strategies of designing. From the standpoint of control theory the task of constructing a minimal-on-time algorithm of designing can be formulated as the problem of finding the transitive process of a dynamic system with the minimal transition time. The main tool in this search is the vector of control U, which allows to change the structure of the functions $f_i(X,U)$, and according to it [35], can change the time of transition. For this purpose, we need guarantee the maximum rate of decrease of the Lyapunov function, i.e. the maximum absolute value of the derivative $\dot{V}$. Typically, this derivative is not positive for a stable process. As shown in the previous section we can state the close relationship between the CPU time of designing and properties of a Lyapunov function of the design process. As shown in the previous section we can state the close relationship between the CPU time of design and properties of a Lyapunov function of the design process. To analyze the process of designing circuits let us introduce a new special function, which is the relative time derivative of Lyapunov function $W = \dot{V}/V$. In this case, we can compare different strategies of designing by analyzing the behavior of the function W(t) during the process of designing, that is, over time, and select the most promising ones, in terms of a minimum of CPU time. Following examples were analyzed based on the continuous form of the formulation of process of designing. Results of the analysis of the behavior of function *W(t)* are given below.

Consider the behavior of the functions *V(t)* and *W(t)* at designing of three-stage transistor amplifier shown in Figure 5. Full structural basis design strategies include 128 different strategies. The behavior of *V (t)* and *W (t)* for some of the strategies of this basis is shown in Figure 20. Results of the analysis of these strategies are presented in Table IX.

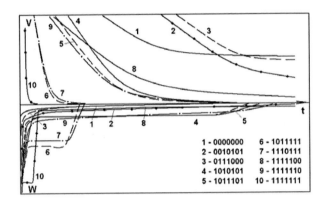

Figure 20. Behavior of the functions *V(t)* and *W(t)* during the design process for certain strategies of the structural basis for the three-stage transistor amplifier.

Table IX. Results of the analysis of design strategies of the structural basis for the three-stage transistor amplifier

N	Control vector	Iterations number	CPU time (sec)
1	(0 0 0 0 0 0 0)	2161512	3019.47
2	(0 0 1 0 1 0 1)	275580	314.94
3	(0 1 1 1 0 0 0)	1075433	1042.05
4	(1 0 1 0 1 0 1)	102510	50.54
5	(1 0 1 1 1 0 1)	107541	43.66
6	(1 0 1 1 1 1 1)	38751	12.81
7	(1 1 1 0 1 1 1)	43387	13.91
8	(1 1 1 1 1 0 0)	160900	97.32
9	(1 1 1 1 1 1 0)	117828	53.72
10	(1 1 1 1 1 1 1)	52651	4.6

Analysis of this example confirms the already discovered earlier dependency between the behavior of the function *W(t)* at the initial stage of the design process and a CPU time of designing. The fastest strategy 10 having the smallest value of the CPU time also corresponds to the largest value of the function *W(t)* at the initial design phase, as shown in Figure 20. Strategies 6 and 7 correspond to smaller values of the function *W(t)* and greater values of processor design time. However, they are much more optimal in time than other strategies. Following in descending order of magnitude $W(t)$ (the degree of increase in CPU time) are the strategies 4 and 5. The remaining strategies, listed in the table have small values of the function $W(t)$ and the greater values of the design-time. In this case, the traditional strategy 1 in 655 times slower than the modified traditional strategy of designing 10.

Generalizing these results, we can conclude the following: the behavior of the relative temporal derivative of the Lyapunov function of the design process $W = \dot{V}/V$ at an early stage of the design process may well predict the total relative time of a strategy of designing. This means that to compare the total CPU time for different design strategies is not necessary to design to completion. It suffices to compare the behavior of $W(t)$ at the initial design stage to determine the strategy with the least CPU time. Greater absolute value of W leads to a smaller total processing time. This property of W suggests that the structure of the optimal time algorithm design should be based on the behavior of this function.

Based on the above analysis we can conclude that the Lyapunov function of the process of designing and function of its derivatives, in particular regarding the time derivative of Lyapunov function can be quite informative source for finding strategies that have small CPU time. Strategies that have the greatest absolute value relative of the time derivative of Lyapunov function in the initial phase of the trajectory of designing are also have the minimal CPU time. This property may provide a basis for the constructing an optimal in time algorithm.

In accordance with the theory of Lyapunov's direct method, information about the stability of the trajectory and in our case and CPU time for optimization of the circuit are related to the time derivative of Lyapunov function. In the current study, in [41, 42] shown that the Lyapunov function of the designing process and its time derivative can be quite informative source for search most perspective strategies of designing. We showed the close correlation between the processor time of designing and behavior of the Lyapunov function of the design process and its derivative. By changing the vector of control U, we can change the CPU time of circuit designing and analyze the behavior of the Lyapunov function and its derivative. This allows us to relate the structure of the control vector with the behavior of the Lyapunov function and with CPU time. The vector of control U is a main tool in the search of strategy with minimal CPU time. This vector allows us a changing of the structure of functions F, and in accordance with this [43, 44] change the time of the transitive process of a dynamic system. Future analysis can provide more information about the properties of the optimal strategy and the structure of optimal algorithm of designing.

Conclusion

Enhancing the effectiveness of the design of analog IP, and especially the circuitry design is one of the priorities of micro and nano electronics. From this task depends largely the progress in the development of new electronic devices and systems, modern information technologies and advanced positions in the market development of these products. To improve the efficiency of circuit design is important to bring new ideas that arise in the allied sciences, and especially in mathematics. One of the major steps in this direction is associated with a more general formalization of the process of circuit design. Formalization of the process of designing of electronic circuits on the conditions of non-compliance with the laws of Kirchhoff leads to a generalization of the design process and the emergence of a number of different strategies of designing. In this case, the conventional approach to the design, or that the same traditional strategy of optimizing electronic circuits, is only one possible strategy in this set.

It is shown that the potential gain in time, which provides optimal or quasi-optimal strategy of designing compared with the traditional approach, will increase when increase the size and complexity of electronic circuits. However, to realize this potential gain, it is necessary to construct an optimal algorithm. This problem can be solved by bringing the ideas and methods of optimal control theory. Introduction of a special control vector in the formulation of the process of circuit's optimization enables us to generalize the task of designing and formulating it in terms of optimal control theory. In this case, the control vector carries the redistribution of CPU time between the two main blocks of the algorithm of circuit designing, an analysis of circuit and a block of parametric optimization. It is shown that such a transformation can improve the process of optimization of the circuit of hundreds or thousands of times when we constructed an optimal structure of the vector of control.

On the basis of generalization of the circuit designing a new effect was detected for further acceleration of the optimization process. This effect appears only in the context of a generalized design methodology and is the first necessary element for the construction of quasi-optimal in time algorithm for designing electronic systems. The second necessary element for constructing quasi-optimal algorithm is choosing the right sequence of strategies that make up the algorithm. Construction of such a sequence was made possible through two strategies, the traditional strategy of designing and the modified traditional strategy of designing. The third element of quasi-optimal algorithm is the choice of switching points from one strategy of designing to another. These points can be found on basis of analysis of the fundamental properties of the designing process, which is presented as a controllable dynamic system.

It is shown that the task of analyzing the temporal characteristics of different strategies of designing is related to the more general problem of stability and convergence of designing strategies. The analysis conducted on the basis of Lyapunov's direct method, by studying the behavior of the Lyapunov function of the design process revealed a significant correlation between the CPU time and the characteristics of a Lyapunov function. Revealed that to compare the total CPU time for different strategies of designing is not necessary the design till final point. It suffices to compare the behavior of the Lyapunov function and its time derivative at the initial design stage to determine the strategy with the least CPU time. In this case, a large absolute value of the normalized derivative of the Lyapunov function leads to less total CPU time. This property suggests that the structure of the optimal in time algorithm of designing should be based on the behavior of this function.

Acknowledgments

We acknowledge the support from Autonomous University of Puebla, under the project VIEP 166/EXC/11-G.

References

[1] Bunch, J.R. & Rose, D.J. (Eds.). (1976). *Sparse Matrix Computations*, Acad. Press : N.Y.

[2] Osterby, O., & Zlatev, Z. (1983). *Direct Methods for Sparse Matrices*, Springer-Verlag: N.Y.

[3] Wu, F.F. (1976). Solution of Large-Scale Networks by Tearing, *IEEE Trans Circ Sys*, vol 23 (12), 706-713.

[4] Sangiovanni-Vincentelli, A., Chen, L.K., & Chua, L.O. (1977). An Efficient Cluster Algorithm for Tearing Large-Scale Networks, *IEEE Trans Circ Sys*, vol 24 (12), 709-717.

[5] Rabat, N., Ruehli, A.E. Mahoney G.W. & Coleman, J.J. (1985). A Survey of Macromodeling, *Proc. of the IEEE Int. Symp. Circuits Systems (ISCAS'85)*, 139-143.

[6] Ruehli, A.E., Sangiovanni-Vincentelli, A. & Rabbat, G. (1982). Time Analysis of Large-Scale Circuits Containing One-Way Macromodels, *IEEE Trans Circ Sys*, vol 29 (3), 185-191.

[7] Ruehli A.E., & Ditlow, G. (1983). Circuit Analysis, Logic Simulation and Design Verification for VLSI, *Proc. IEEE*, vol 71 (1), 36-68.

[8] George, A. (1984). On Block Elimination for Sparse Linear Systems, *SIAM J. Numer. Anal.* vol 11 (3), 585-603.

[9] Fletcher, R. (1980). *Practical Methods of Optimization*, John Wiley and Sons: N.Y.

[10] Gill, P.E., Murray, W. & Wright, M.H. (1981). *Practical Optimization*, Academic Press: London.

[11] Brayton R.K., Hachtel G.D., & Sangiovanni-Vincentelli A.L. (1981). A survey of optimization techniques for integrated-circuit design, *Proc. IEEE*, vol 69, 1334-1362.

[12] Ruehli, A.E. (1987). *Circuit Analysis, Simulation and Design*, vol 3, Elsevier Science Publishers: Amsterdam.

[13] Massara, R.E. (1991). *Optimization Methods in Electronic Circuit Design*, Longman Scientific & Technical: Harlow.

[14] Kashirskiy, I.S. (1976). General Optimization Methods, *Izvest. VUZ USSR - Radioelectronica*, vol 19 (6), 21-25.

[15] Kashirsky I.S. & Trokhimenko, Y.K. (1979). *The Generalized Optimization of Electronic Circuits*, Tekhnika: Kiev.

[16] Rizzoli, V., Costanzo A. & Cecchetti, C. (1990). Numerical optimization of broadband nonlinear microwave circuits, *IEEE MTT-S Int. Symp.*, vol 1, 335-338.

[17] Ochotta, E.S., Rutenbar R.A. & Carley, L.R. (1996). Synthesis of High-Performance Analog Circuits in ASTRX/OBLX, *IEEE Trans on CAD*, vol 15 (3), 273-294.

[18] Ochotta, E., Mukherjee, T., Rutenbar, R. & Carley, L.R. (1998). *Practical Synthesis of High-Performance Analog Circuits*, Kluwer: Norwell, MA.

[19] Zemliak A. (2001). One Approach to Analog System Design Problem Formulation, *Proc. 2001 IEEE Int. Sym. on Quality Electronic Design (ISQED2001)*, 273-278.

[20] Zemliak A. (2001). System Design Problem Formulation by Control Theory, *Proc. IEEE Int. Sym. on Circuits and Systems (ISCAS2001)*, vol 5, 5-8.

[21] Zemliak A.M. (2001). Analog System Design Problem Formulation by Optimum Control Theory, *IEICE Trans on Fundam of Electr Commun and Computer Sciences*, vol E84-A (8), 2029-2041.

[22] Zemliak A. (2002). Novel Approach to the Time-Optimal System Design Methodology, *WSEAS Trans Sys*, vol 1 (2), 177-184.

[23] Zemliak, A.M. (2004). Analog Circuits Design by the Methods of Control Theory, Part 1, Theory, *Izv VUZ Radioelektronika,* vol 47 (5), 18-28.

[24] Zemliak, A.M. (2004). Analog Circuits Design by the Methods of Control Theory, Part 2, Numerical Results, *Izv VUZ Radioelektronika,* vol 47 (6), 65-71.

[25] Pontryagin, L.S., Boltyanskii, V.G., Gamkrelidze, R.V. & Mishchenko, E.F. (1962). *The Mathematical Theory of Optimal Processes,* Interscience Publishers, Inc.: New York.

[26] Rosen, J.B. (1966). Iterative Solution of Nonlinear Optimal Control Problems, *J. SIAM, Control Series A*, 223-244.

[27] Fedorenko, R.P. (1978). *Approximate Solution of Optimal Control Problems*, Nauka: Moscow.

[28] Sepulchre, R., Jankovic, M. & Kokotovic, P.V. (1997). *Constructive Nonlinear Control*, Springer-Verlag: New York, N.Y.

[29] Slotine J.E., & Li, W. (1991). Applied Nonlinear Control, Englewood Cliffs Prentice-Hall: NJ.

[30] Krylov, I.A. & Chernousko, F.L. (1972). Consecutive Approximation Algorithm for Optimal Control Problems, *J. of Numer. Math. and Math. Pfysics*, vol 12 (1), 14-34.

[31] Pytlak, R. (1999). *Numerical Methods for Optimal Control Problems with State Constraints*, Springer-Verlag: Berlin.

[32] Massobrio G. & Antognetti P. (1993). *Semiconductor Device Modeling with SPICE*, Mc. Graw-Hill, Inc.: N.Y.

[33] Zemliak, A.M. (2002). Acceleration Effect of System Design Process, *IEICE Trans on Fundam of Electr Commun and Computer Sciences, vol E85-A* (7), 1751-1759.

[34] Zemliak, A.M. (2004). Analysis of the Acceleration Effect of the Analog Circuits Design by the Methods of Control Theory, *Izv VUZ Radioelectronica,* vol 47 (7), 52-59.

[35] Zemliak, A.M. (2004). Analysis of the Initial Point Structure and the Design Trajectories for Analog Circuit Design, *Izv VUZ Radioelectronica,* vol 47 (12), 3-11.

[36] Zemliak A.M. (2007). Separatrix Conception for Trajectory Analysis of Analog Networks Design in Minimal Time, *IEICE Trans on Fundam of Electr Commun and Computer Sciences, vol E90-A* (8), 1707-1712.

[37] Barbashin, E.A. (1967). *Introduction to the Stability Theory*, Nauka: Moscow.

[38] Rouche, N., Habets, P. & Laloy, M. (1977). *Stability Theory by Liapunov's Direct Method*, Springer-Verlag: N.Y.

[39] Zemliak, A.M. (2006). Analog System Design Problem as Controllable Dynamic Process, *Nonlinear World,* vol 4 (11), 609-618.

[40] Zemliak, A.M. & Markina, T.M. (2009). Electronic circuit optimization as a controllable dynamic process, *Management Inform Automatic Sys and Dev,* vol 146, 62-69.

[41] Zemliak, A.M. (2009). A structure of time minimal strategy of analog circuits optimization, *Radioelectr and Communic Sys,* vol 52 (1), 32-37.

[42] Zemliak, A.M. (2009). Analysis of the control vector structure in analog networks design, *Radioelectr and Communic Sys,* vol 52 (10), 530-536.

[43] Zemliak A., Torres, M. (2009). On Structure of the Control Vector for Minimal-Time Networks Design Strategy, *WSEAS Trans Circ Sys,* vol 8 (12), 905-915.

[44] Zemliak A.M. (2009). Structure of the Control Vector for the Minimal-Time Circuit Design Algorithm, *Proc Int. IEEE Conf. devoted to the 150-anniversary of A.S. Popov (EUROCON2009)*, 2046-2051.

In: Analog Circuits: Applications, Design and Performance
Editor: Esteban Tlelo-Cuautle

ISBN: 978-1-61324-355-8
© 2012 Nova Science Publishers, Inc.

Chapter 8

A TECHNOLOGY-AWARE OPTIMIZATION OF RF INTEGRATED INDUCTORS

P. Pereira[1], A. Sallem[2], M.H. Fino[1,], M. Fakhfakh[2] and F. Coito[1]*

[1] FCT – Universidade Nova de Lisboa, Portugal
[2] University of Sfax, Tunisia

Abstract

This Chapter presents the optimal design of radio-frequency integrated spiral inductors. The basic idea is to generate an analytical model to characterize integrated inductors based on the double π-model, and offer to the designer an approach to determine the inductor layout parameters. Particle Swarm Optimization technique is used to generate optimal values of parameters of the developed models. Viability of the proposed models is highlighted via comparison with *ASITIC* simulation results.

Introduction

Modern CMOS processes have reached a maturity that makes them suitable for implementing radio-frequency (RF) devices fulfilling the requirements of the wireless standards working at the GHz frequency range. Due to their low cost and ease of process integration, on-chip spiral inductors are widely used in analog blocks of RF integrated circuits, such as voltage-controlled oscillators (VCOs), low-noise amplifiers (LNAs) and passive-element filters [1].

Notwithstanding the widespread use of integrated inductors, their design is still a quite challenging task due to the complexity of design options that must be adopted aiming at minimizing the inductors high frequency poor performance and their subsequent impact on the circuit efficacy. This poor performance is due to the large effect the technology parasitics have on the usually required small value of inductance. As a result, significant effort has been employed in investigating silicon planar inductors, their associated models and methods of

---

*Corresponding author

improving their performance. Modern fabrication costs are putting pressure on designers to produce ready to market designs in less iteration. This reduces the opportunity for using a specific process fine-tune *S*-parameter based [2] or empirical [3] models. On the other hand accurate *SPICE* level inductor models are restricted to an insufficient number of geometries forcing designers of custom inductors to a time consuming process of electromagnetic (EM) simulation. Furthermore, the use of either of the previously mentioned approaches, although producing accurate results, does not give the designer a qualitative insight into which parameters should be altered in order to improve the inductor behavior. Thus, scalable and physically-based models are of vital importance for the designer to make decisions regarding the improvement of inductor behavior. Furthermore, such models also allow to optimize the inductor within the circuit design process. Previous attempts at modelling integrated inductors have concentrated on the simple π-model [4] which does not account for all the high frequency and parasitic effects. The necessity for overcoming these limitations has motivated the development of the double π-model [5]. In [6] a tool for the automatic generation of an analytical model for characterizing integrated inductors based on the double π-model was presented. Since designing a spiral inductor involves the determination of multiple correlated variables, optimization based techniques have been proposed [7]- [12]. The double π-model is included in the optimization tool presented in [13] allowing the design of spiral inductors for operating frequencies beyond 1 GHz. Incorporating an accurate model and taking into account technology constraints, makes this tool a good instrument for a first approach on inductors' design. The accuracy of the results obtained with the tool was compared with results reported in the bibliography.

In this Chapter, we start by presenting a brief introduction to integrated inductor design variables, as well as the analythical model adopted for characterizing the inductor behavior [6]. After justifying the need for optimization techniques in the design of integrated inductors an introduction to optimization-based design of circuits is presented. Particular enphasys is given to Particle Swarm Optimization heuristics. The methodology for the optimization based design of RF inductors is described and then, working examples illustrating both mono-objective and multiobjective optimization of inductors are presented. Results are first compared to the well known NSGA-II then against *ASITIC* simulations. Finally conclusions are offered.

Spiral Inductors

Integrated spiral inductors can be implemented in silicon substrate by using multilevel interconnects provided by the silicon fabrication process. The design of an integrated inductor consists in the determination of geometrical parameters, such as the number of sides (square, hexagonal, octagonal), number of turns, *n*, the turn spacing, *s*, the turn width, *w,* and the outer diameter, *dout*, as illustrated in Figure 1

A Technology-Aware Optimization of RF Integrated Inductors

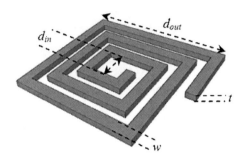

Figure 1. Layout parameters for a square inductor.

For evaluating an inductor design, several characteristics, such as the maximum quality factor, Q, the frequency for the maximum quality factor, the inductance value, the self-resonance frequency, f_{SR}, or the area, may be considered. The choice for some of these characteristics in detriment of the remaining depends on the circuit into which the inductor is to be used.

During the last years several equivalent-circuit inductor models have been proposed as illustrated in Figure 2 [18].

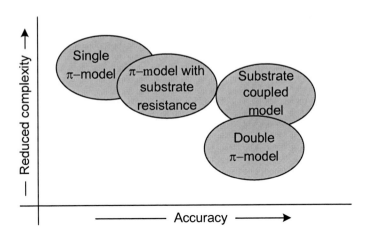

Figure 2. Advantages of the double-π model over the single π-model.

The double-π model has been proposed with the aim of overcoming the lack of accuracy of results obtained with the single π-model for higher frequencies. This limitation stems from the fact that analytical expressions for the evaluation of the π-model lumped elements do not take into account some high frequency effects, such as the skin and the proximity effects [5].

The double π-model is represented in Figure 3. This model uses a wide range of equations for evaluating the inductor model lumped element values. Some of those elements such as L_0 and L_p, depend on the inductance value at low frequencies, L_{dc}, where parasitics do not affect the inductor behaviour. For the estimation of the spiral inductance, L_{dc}, many inductance models can be found in literature, such as the Modified Wheeler Formula or the Greenhouse Approximation of Grover [14].

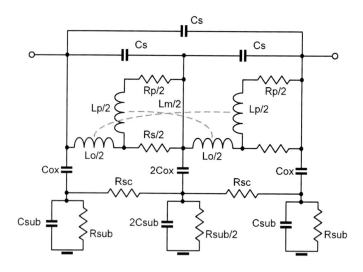

Figure 3. Inductor double π-model.

For the evaluation of the model parameters a classification into *DC* inductor parameters, Substrate *Network* and *Ladder* Circuit elements, is considered.

- **DC Inductor parameters**

The DC inductor parameters consist of R_{dc}, L_{dc} and the capacitances C_c and C_s and C_{ox}. The DC resistance is obtained with

$$R_{dc} = R_{sheet} \frac{l}{w} \qquad (1)$$

where R_{sheet} is the metal sheet resistance per area, l is total metal length and w is the metal with. Regarding the capacitances involved, three of them are considered; metal-to-metal (C_s and C_c) and metal-to-substrate (C_{ox}) capacitance. The crossover capacitance, C_s, appears between the spiral and the underpass connecting the inner turn to the outside of the spiral inductor. For the evaluation of this capacitance, all overlap capacitances are considered [15].

$$C_s = n_c w^2 \frac{\varepsilon_{ox}}{t_{oxM1-M2}} \qquad (2)$$

where ε_{ox} is the oxide permittivity, n_c is the number of overlaps and $t_{oxM1-M2}$ is the oxide thickness between the spiral upper and lower metal. The metal-to-metal capacitance C_c occurs due to the proximity of inductor tracks, whereas C_{ox} accounts for the metal-to-substrate capacitance. Both capacitances are modelled as the summation of three rational functions which simulate three flux components, and are obtained by [16].

$$C_c = 2 l \varepsilon_{ox} \left(1.144 \frac{t_{metal}}{s} \left(\frac{t_{ox}}{t_{ox}+2.059 s} \right)^{0.0944} + .7 + 1.158 \left(\frac{w}{w+1.874 s} \right)^{0.1612} \right) \qquad (3)$$

A Technology-Aware Optimization of RF Integrated Inductors

$$C_{ox} = l\,\varepsilon_{ox}\left(\frac{w}{t_{ox}} + 2.217\left(\frac{s}{0.702\,t_{ox}+s}\right)^{3.193} + 1.1\left(\frac{t_{metal}}{t_{metal}+4.532\,t_{ox}}\right)^{0.1204}\right) \tag{4}$$

l is the spiral length.

- **Substrate *Network***

Concerning the substrate elements, there's the need to model the ohmic losses in the conductive silicon substrate, which are encapsulated in R_{sub} and C_{sub}. A third element is R_{sc} which represents the electric coupling between lines through the conductive substrate, and given by

$$R_{sc} = \frac{1.5\,\rho_{substrate}\,N\,P}{l\,t_{substrate}} \tag{5}$$

where N is the number of turns and P is the metal line pitch (metal width plus the metal spacing from edge to edge).

As a way to predict the frequency influence in substrate losses, R_{sub} and C_{sub} should be frequency dependent equations given by

$$C_{sub} = l\,\frac{\omega^2\,C_1 C_2\,(C_1 + C_2) + C_1 G_s^{\,2}}{G_s^{\,2} + \omega^2\,(C_1 + C_2)^2} \tag{6}$$

$$R_{sub} = l\,\frac{\omega^2\,C_1 G_s}{G_s^{\,2} + \omega^2\,(C_1 + C_2)^2} \tag{7}$$

where, C_1 and C_2 can be obtained through expressions in [17].

- ***Ladder* Circuit Elements**

The set of equations needed for the computation of the extra elements behaviour is

$$R_s = \left(1 + n_r^{-1}\right) R_{dc} \tag{8}$$

$$L_s = \left(1 - 3.57\,\frac{k}{n_r^{1.5}}\right) L_{dc} \tag{9}$$

$$R_p = n_r\,R_s \tag{10}$$

$$Lp = \frac{L_s}{0.315\,n_r} \tag{11}$$

$$L_m = K\,\sqrt{L_s\,L_p} \tag{12}$$

with

$$K = \sqrt{\frac{0.315}{m}} \left(\frac{R_{dc}}{\omega_{crit} L_{dc}} \right) n_r \qquad (12.a)$$

and

$$\omega_{crit} = \frac{3.1}{\mu_0} \frac{P}{\omega^2} R_{sheet} \qquad (12.b)$$

As we can easily infer from the equations above, designing a spiral inductor consists in the evaluation of a set of highly correlated geometrical layout parameters. Furthermore these parameters are subject to constraints arising from the technology process used. To efficiently cope with the characteristics mentioned, optimization techniques must be used for designing integrated inductors.

Optimization Based Design

Typically, the analog, mixed signal and radio-frequency (AMS/RF) design problems consist of formulating the requirements in terms of bounds on performance functions. Analytical equations that predict these performance functions are first expressed in terms of device model parameters and then replaced by their symbolic expressions at the design variables level, such as the MOS transistor geometries[19].

Optimizing the design of a circuit consists in finding the sizing variable set that optimizes performance function(s), meets imposed specifications, and satisfies inherent constraints (working mode of the transistors, technology limitations, etc.) [20].

Actually, the general optimization problem definition can be formatted as follows:

$$\begin{vmatrix} Minimize\ f(\vec{x}); & f(\vec{x}) \in R \\ such\ that: \\ g(\vec{x}) \le 0; & \vec{g}(\vec{x}) \in R^m \\ where\ x_{li} \le x_i \le x_{ui}, & i \in [1, p] \end{vmatrix} \qquad (13)$$

m inequality constraints to satisfy,

p parameters to manage.

$\vec{x}_l$ and $\vec{x}_u$ are lower and upper boundaries vectors of the parameters.

Pushed by the progress in technology, the design teams are facing a curve of complexity that grows exponentially, thus traditional design methods that were/are based on the designer experience, the use of statistical-based methods, i.e. trial-and-error approaches involving the use of circuit simulators, and some times the rule-of-thumb, are no longer valid to face and solve such sizing problems. Indeed, these design methods are unstructured, time-consuming, error prone, and requires the attention of a skilled designer [20].

Furthermore, more often than not, circuit optimization problems encompass many objectives; e.g. maximizing the gain, minimizing the noise figure, maximizing the bandwidth, minimizing power consumption, etc. Consequently (13) should be modified as:

$$\begin{vmatrix} \text{Minimize } \vec{f}(\vec{x}); \quad \vec{f}(\vec{x}) \in R^k \\ \text{subject to:} \\ \vec{g}(\vec{x}) \le 0; \quad \vec{g}(\vec{x}) \in R^m \\ \text{where } x_{Li} \le x_i \le x_{Ui}, \quad i \in [1, p] \end{vmatrix} \quad (14)$$

where, k is the number of objectives (≥ 2), m is the number of inequality constraints to satisfy and p parameters to manage.

Figure 4 gives a pictorial view of design optimization approach.

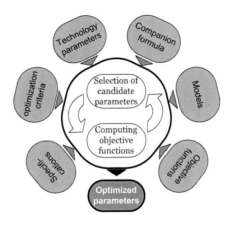

Figure 4. Pictorial view of a design optimization approach [21].

Actually, the literature offers a wide variety of publications dealing with the resolution of this optimization problems and studying in particular the convergence properties of such methods (e.g.[22][24]). It has already proven that the aforementioned classical optimization techniques are by and large not satisfactory. Metaheuristics were proposed and are nowadays widely used to solve such constrained problems [22]. Metaheuristics offer the advantages to be flexible to be modified and adapted to suit the problem requirements [22][24]. Even though they don't guarantee convergence to the global optimal solution, a fact due to their stochastic nature, they give, within an acceptable computing time, a good approximation of it. In fact, Metaheuristics can be classified into two categories [24]:

• The population based techniques, such as:

- Evolutionary Algorithms (EA) [25][26]: Genetic Algorithms (GA) [27], etc., and

- Swarm Intelligence Techniques (SI) [28]: Particle Swarm Optimization (PSO) [29][30], Ant Colony Optimization (ACO) [31], Bacterial Foraging Optimization (BFO) [32],etc.

and
• The single solution based techniques, such as Simulated Annealing (SA) [33], Tabu Search (TS) [34][35]…

Some among the aforementioned optimization techniques were adopted to optimize performances of AMS/RF circuits, see for instance [13],[36]-[38]. Basically, the above mentioned optimisation techniques were proposed to deal with monobjective optimization problems. In order to handle multiobjective design and sizing problems, designers commonly use the weighted sum approach [39]. Preferences are taken into account by assigning several weightings, for each objective function $f_i(x)$. A weighted sum approach transforms a multi-objective optimization problem in a single-objective optimization problem $F(x)$. The so obtained monobjective function can be written as:

$$F(\vec{x}) = \Sigma_{i=1}^{k} \omega_i f_i(\vec{x})$$

(15)

where $\omega_i, i \in [1,k]$, are weighting coefficients. It is usually assumed that $\Sigma_{i=1}^{k} \omega_i = 1$.

However, it has already been proven that such technique suffers from many drawbacks, mainly the fact that it only allows the generation of a unique 'optimal' solution. It is to be highlighted that varying the weighting coefficients cannot allow generating all the non-dominated optimal solutions [40].

Indeed, as it was introduced above, most design optimisation problems comprise more that one objective. These objectives have contradictory behaviours, i.e. optimizing one performance leads to the degradation of another one. The dominance process was introduced and thus the Pareto dominance[1] [41].

Most famous such techniques were adapted to deal with multiobjective problems, for instance Multiobjective Genetic Algorithm MOGA [42], Non-dominated Sorting Genetic Algorithm NSGA-II [43](derived from GA), Multi Objective Particle swarm optimization MOPSO [39,44] (from PSO), Multiobjective Simulated Annealing MOSA [45] (from SA) etc. The basic concept consists of generating (all) the non-dominated solutions forming the Pareto front [23],[41],[46].

Finally, it is to be highlighted that classical metaheuristics cannot handle inequality constraints, so the constrained optimization problem is transformed into an unconstrained one by minimizing the following function:

$$\tilde{f}(x) = f(x) + \gamma(d(x, Fr))$$

(16)

---

[1] An objective vector $f(x)$ is said to dominate another objective vector $\vec{f}(y)$ (i.e. $\vec{f}(x) \succ \vec{f}(y)$) if no component of $\vec{f}(x)$ is smaller than the corresponding component of $\vec{f}(y)$ and at least one component is greater.

where $d(x,Fr)$ is a distance metric of the infeasible point to the feasible region Fr. It may simply be zero if no constraint violation occurs and is a positive scalar, otherwise [26],[47]. The definition of this distance metric includes the penalty coefficients that are used to stress the importance of a particular constraint violation over others. γ is a weight coefficient to set a global importance of the constraints with respect to the objective function [47].

Particle Swarm Optimization Techniques

Particle swarm optimization techniques is a global population-based metaheuristic for dealing with problems in which best solution can be represented as a point or a surface in an n-dimensional space. It was first described in 1995 [29]. It belongs to the category of Swarm Intelligence (*SI*) techniques.

SI addresses mechanisms taken from nature by which groups of independent individuals achieve some objectives when operating together [48].

SI systems are typically made up of a population of simple agents interacting locally with one another and with their environment. The agents follow simple rules, and although there is no centralized control structure deducting how individual agents should behave, local interaction between such agents lead to the emergence of complex global behavior [49]. Natural examples of SI include PSO, ACO, Bee Colony Optimization [50] (BCO), and Bacterial Foraging Optimization (BFO) [51].

Actually, PSO is a bio-inspired optimization method that imitates the social conduct of some animals in search of food (fish, birds).

When compared to evolutionary techniques, mainly genetic algorithms (GA), PSO shares some main procedures:

- starting via a random generation of a population (a swarm),
- evaluating fitness values of the population,
- no guarantee to reach the global optimum.

On the other hand, unlike GA, PSO has no evolution operators (cross-over, mutation). Particles update them selves thanks to their internal velocity and their memory.

Further, the PSO information sharing mechanism is different from the one of GAs; regarding GA, chromosomes share information with each other, thus, the whole population moves like a unique group towards the 'optimum'. In PSO, only *gbest* (or *lbest*) gives the information to others [49].

Main advantages of PSO can be summarized as follows:

- it is insensitive to scaling of the design parameters,
- it is simple to be implemented,
- it is derivative free,
- it has few (tuning) parameters,
- it has a very efficient global search mechanism (*vs.* weak local search ability).

In the PSO algorithm, particles fly through the design (parameter) space. During their flights, individuals try to move towards the fittest position[2] known to them and to their informants, i.e. the set of individuals that form the social circle, as follows:

- each particle keeps track of its best coordinates, i.e. coordinates in the variable space associated with the best performance it has achieved so far, called 'pbest',
- the particle also tracks the best value obtained so far by any particle in the social neighbourhood; lbest,
- finally, the particles take also into consideration the best global value of all the swarm, i.e. gbest.

Figure 5 illustrates this principle.

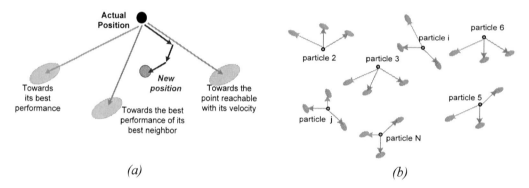

Figure 5. (a) Principle of the movement of a particle, (b) movement of the particles within the swarm (taken from [52]).

Thus, the optimization process updates, at each time step, its velocity and position as follows [53]:

$$\vec{v}_{k+1}^i = \omega_k \vec{v}_k^i + c_1 r_1 (\vec{p}_k^i - \vec{x}_k^i) + c_2 r_2 (\vec{p}_k^g - \vec{x}_k^i) \qquad (17)$$

$$\vec{x}_{k+1}^i = \vec{x}_k^i + \vec{v}_{k+1}^i \qquad (18)$$

ω_k is an inertia weight, generally it is in training progress using a decreasing inertia function (see (19)), in order to lessen the influence of past velocities. c_1 is a cognition learning rate, c_2 is a social learning rate, p_k^i is the best remembered individual position, p_k^g is the best remembered swarm position, $r_{1,2}$ are random parameters between 0 and 1.

$$\omega_k = \omega_k^{max} - \frac{\omega_k^{max} - \omega_k^{min}}{max\_nb\_iter} current\_iter \qquad (19)$$

---

[2] The particle position represents a candidate solution to the optimization problem at hand.

A Technology-Aware Optimization of RF Integrated Inductors

ω_k^{max} is the initial weight, ω_k^{min} is the final weight, max_nb_iter is the maximum iteration number, and $current\_iter$ is the current iteration number.

The basic pseudo-code of PSO is:

Initialization
   Set constants
For each particle
   Randomly initialize position $\vec{x}_0^i$

   Randomly initialize velocities $\vec{v}_0^i$

End For
Do
   For each particle
   Evaluate the fitness value of all particles f_k^i using design space coordinates

   $\vec{x}_k^i$
   If the fitness value is better than *pbest*
     Set current value as the new *pbest*
   End If
   End For
   Choose the particle with the best fitness value of all the particles as the *gbest*
   For each particle
   Compute particle velocity (according to (5))
   Compute particle position (according to (6))
   End For
While the stop criteria is not met.

Algorithm 1. Pseudo code for PSO.

Figure 6 illustrates the flowchart of the monobjective PSO:

PSO was adapted to be able to deal with multiobjective optimization problems. The basic idea consists of the use of an archive, in which each particle deposits its 'flight' experience at each running cycle. This archive stores all the non-dominated solutions found during the optimization process thus forming and generating the Pareto Front.

In order to avoid excessive growing of this memory, its size is fixed. This implies to define some rules for the update of the archive. Algorithm 1 depicts the pseudo-code of these rules.

The flowchart presented in Figure 7 depicts the MOPSO [44],[52].

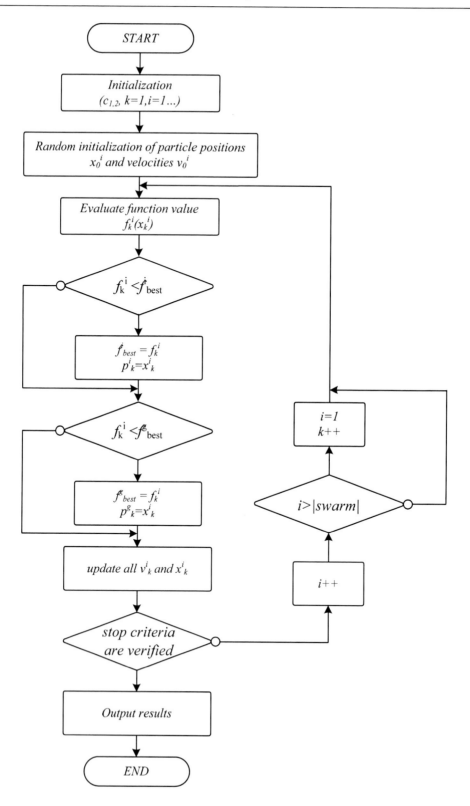

Figure 6. The basic PSO algorithm.

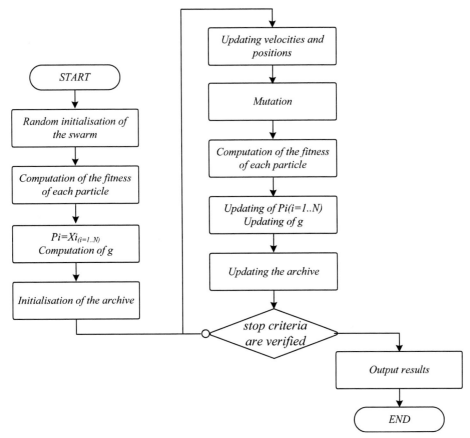

Figure 7. Flowchart of a MOPSO.

Optimization Design of Spiral Inductors

The methodology adopted aims to achieve a technology/topology-aware solution, where constraints are mandatory. Regarding the number of turns, *n*, it was considered to have a minimum value of 1.5.

Due to the inductor model considered the shape of the inductor must account for square, hexagonal and octagonal topologies. On what concerns the technological constraints, minimum values for the track width, *w*, for the track-to-track spacing, *s*, and for in input diameter, d_{in}, must be defined. Technology-depend minimum increment values for these layout parameters must also be considered. Finally the correlation between the layout parameters defined by (21) is considered, as a way of including heuristic design rules for reducing the parasitic phenomena due to the proximity effect [54].

$$0.2 < d_{in}/d_{out} < 0.8, \quad d_{in} > 5W. \tag{20}$$

If we define Cost(n,d_{out},w,s) and L(n,d_{out},w,s) as the cost function and the inductance of the spiral, and L_{exp} and δ the targeting inductance value and the tolerance allowed for the inductance to deviate from the targeting value, the optimization problem is formulated as the minimization of Cost(n,d$_{out}$,w,s) subject to:

$$(1-\delta)L_{exp} \leq L(n,d_{out},w,s) \leq (1+\delta)L_{exp}$$
$$w \in [w_{min}:w_{max}]$$
$$d \in [d_{min}:d_{max}]$$
$$n \in [n_{min}:n_{max}]$$
$$N \in [2,4,8]$$

(21)

Concerning the cost function different scenarios are available yielding the minimization of the device area, d_{out} which expression is given by (22), or the maximization of the quality factor, for a predefined frequency of operation. For the double pi-model, the expression of the quality factor is obtained according to the methodology in [6]. For this reason, we will discuss in the next section two kinds of optimization problems: mono-objective and multi-objective optimization problems.

$$d_{out} = d_{in} + 2.n.w + 2.(n-1).s \qquad (22)$$

where, n is the number of turns, s is the track to track distance, and w is the track width.

It should be emphasized thar the optimization methods used are inherently continuous, i.e., design variable will be considered as continuous. Due to technology constrains only discrete values may be considered as design solutions. To solve this problem a post-processing procedure is to be applied to the solutions found, where the design variable values are rounded to the nearest value allowed fdor the technology used.

Working Examples

It is important to put the stress on the fact that the PSO algorithm can be used for both mono-objective and multi-objective optimization problems.

- *Monobjective problem:*

We applied the monobjective PSO algorithm and we compare it with SA, to perform optimization of spiral inductor double π-model on Silicon technologies.

The design of both 5.0 and 6.0nH inductor for an operation frequency of 0.7 and 1.0 GHz for UMC130 technology is addressed. Technological and physical parameters shown in Table I, as well as layout constraints, given in Table II, were taken into account in the optimization design. In addition a minimum space between tracks (*s*) of 2.5 μm and a maximum output diameter of 200 μm were assumed.

A Technology-Aware Optimization of RF Integrated Inductors

Table I. Physical Parameters of Inductor Design

Parameter	Value
Metal Thickness	2.8 μm
Space betwwen turns	2.5 μm
Sheet Resistance	10 mΩ/square
Oxide Thickness	5.42 μm
Oxide Thickness between spiral and underpass	0.26 μm
Oxide Permittivity (ε_r)	4
Susbtrate Thickness	700 μm
Susbtrate Permittivity (ε_r)	11.9
Susbtrate Resistivity	28 Ω.cm

Table II. Design Constraints

Parameter	min	Max
w (μm)	5.0	50.0
d_{in} (μm)	20.0	250.0
n	1.5	15.5

The objective function to be optimized is the inductor quality factor Q (to be maximized). Table III and IV give the algorithms' parameters:

Table III. The PSO algorithm parameters

Swarm size	Numbers of Iteration	ω	c_1	c_2
40	100	0.4	1	1

Table IV. The SA algorithm parameters

Initial temperature	Stopping temperature	Cooling schedule	Numbers of Iteration
$1e^5$	$1e^{-08}$	0.9	1000

Figures 8 and 9 represents the optimization results corresponding to the PSO (8.a, 8.b) and SA (9.a, 9.b) for the optimization of an inductor of 5.0 nH and an inductor of 6 nH for an operating frequency of 0.7 GHz and 1 GHz, respectively, each of them for the maximization of the inductor quality factor Q. We notice that the problem of maximizing Q is transformed into the minimization of *(-Q)*.

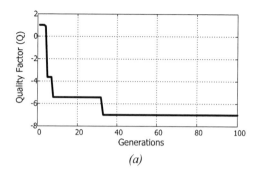 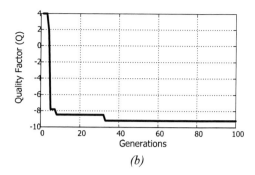

Figure 8. PSO optimization results (Q vs. number of generations): *(a)* 5 nH-inductor operating at 0.7 GHz, *(b)* 6 nH-inductor operating at 1 GHz.

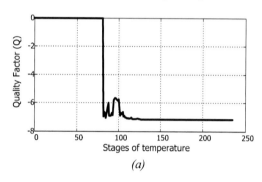 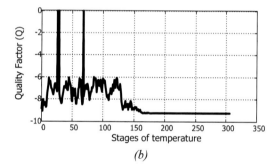

Figure 9. SA optimization results (Q vs. stages of temperature): *(a)* 5 nH-inductor operating at 0.7 GHz, *(b)* 6 nH-inductor operating at 1 GHz.

Table V gives parameter values (inductor sizes) obtained using both algorithms. Besides, it gives theoretical and simulation results values of the objective function. The validity of these results was checked against *ASITIC* simulations.

Table V. Optimization and simulation results

$L = 5nH$, $Freq. = 0.7\ GHz$

Algo.	w (μm)	d_{in} (μm)	n	$d_{out\text{-}Op}$ (μm)	L_{Op} (nH)	Q_{Op}	L_{ASITIC} (nH)	Q_{ASITIC}	Time (sec)
PSO	9.96	53.61	5.9	195.94	5.08	6.98	4.70	5.85	10.05
SA	10.44	62.59	5.5	200	4.87	7.01	4.83	6.2	83.38

$L = 6nH$, $Freq. = 1.0\ GHz$

Algo.	w (μm)	d_{in} (μm)	n	$d_{out\text{-}Op}$ (μm)	L_{Op} (nH)	Q_{Op}	L_{ASITIC} (nH)	Q_{ASITIC}	Time (sec)
PSO	9.00	63.78	5.97	196.16	6.03	9.20	5.25	8.20	4.48
SA	9.18	76.46	5.5	200	5.85	9.29	5.60	8.59	100.18

It is to be highlighted that, even though the two tested heuristics converge to very similar optimal solutions, the PSO algorithm presents the potentiality to converge rapidly to the global optimum and within a reduced number of parameters to manage. *ASITIC* simulation results are also given in the tables above and fully demonstrate the validity of the obtained results.

- *Multi-objective problem:*

The aim consists of optimizing two conflicting performances: maximizing the Quality factor, Q, and minimizing the device area, d_{out} by generating the trade-off surface (Pareto front). We give optimization results and present comparison with results obtained using the well known NSGA-II technique.

Figure 10 represents Pareto fronts obtained using MOPSO and NSGA-II where it can be noticed that the better distribution given by MOPSO, when compared to the NSGA-II one.

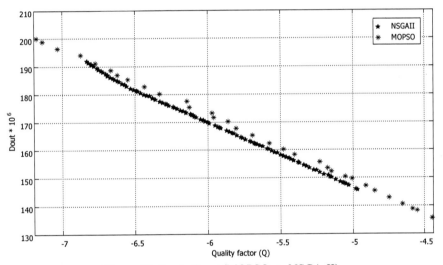

Figure 10. Pareto Front (MOPSO *vs.* NSGA-II).

Table VI presents optimal performances of two solutions taken from the archive. They correspond to edges of the Pareto boarder: solutions giving the maximum quality factor (solution1) and the minimum D_{out} (solution2), respectively.

Table VI. Optimization and simulation results (MOPSO *vs.* NSGA-II)

Solution 1, $L = 4nH$, freq. $= 0.7$ GHz

Algo.	w (µm)	d_{in} (µm)	n	L_{Op} (nH)	$d_{out-max}$ (µm)	Q_{max}	L_{ASITIC} (nH)	Q_{ASITIC}
MOPSO	10.54	61.52	5.5	4.80	200	7.21	4.76	6.19
NSGA-II	10.26	69.20	5	4.23	192.2	6.59	4.16	5.76

Table VI. Continued

Solution 2, L = 4nH, freq. = 0.7 GHz

Algo.	w (µm)	d_{in} (µm)	n	L_{Op} (nH)	$d_{out-min}$ (µm)	Q_{min}	L_{ASITIC} (nH)	Q_{ASITIC}
MOPSO	6.67	39.64	5.5	3.10	135.47	4.44	3.09	3.86
NSGA-II	7.54	41.60	5.5	3.31	147.04	4.97	3.30	4.27

Conclusion

This chapter presented the optimal design of RF integrated spiral inductors, through the use of Partticle Swarm Optimization. The efficiency of the design process is granted by using an analytical model to characterize integrated inductors based on the double π-model. Physically-based equations for the evaluation of the model parameters are considerd, as a way of easily adopting the model to any new technology. Particle Swarm Optimization technique is used to generate optimal values of parameters of the developed models. Several working examples for monobjective and multiobjective optimization were considerd. The viability of the obtained design solutions is highlighted via comparison with *ASITIC* simulation results.

References

[1] Burghartz, J.N *et al* (1998), RF circuit design aspects of spiral inductors on silicon , *IEEE Journal of Solid-State Circuits,* vol 33(12), 2028–2034.

[2] Horng, T.S., Wu, J.M., Yang, L.Q. & Fang, S.T. (2003). A Novel Modified-T Equivalent Circuit for Modeling LTCC Embedded Inductors with a Large Bandwidth. *IEEE Trans. Microwave Theory Tech.*, vol 51(12), 2327–2333.

[3] Scuderi, A., Biondi, T., Ragonese, E. & Palmisano, G. (2003). A ScalableModel for Silicon Spiral Inductors. *IEEE MTT-S Intl. Microwave Symp. Digest*, vol 3, 2117–2120.

[4] Yue, C.P. & Wong, S.S. (2000). Physical Modeling of Spiral Inductors on Silicon. *IEEE Trans. Electron Devices*, vol 47(3), 560–568.

[5] Yu Cao, Groves, R.A., Zamdmer, N.D., Plouchart, J.-O., Wachnik, R.A., Xuejue Huang, King, T.-J. & Chenming Hu. (2003). Frequency-independent equivalent-circuit model for onchip spiral inductors. *IEEE Journal of Solid-State Circuits,* vol 38(3), 419–426.

[6] Pereira, P., Fino, M.H. & Ventim-Neves, M. (2010). Automatic Generation of RF Integrated Inductors Analytical Characterization. *International Workshop on Symbolic and Numerical Methods, Modeling and Applications to Circuit Design (SM2ACD'10)*, 1–4.

[7] Nieuwoudt, A. & Massoud, Y. (2005). Multi-level approach for integrated spiral inductor optimization. *The Design Automation Conference*, 648–651.

[8]   Allstot, D., Choi, K. & Park, J. (2003). *Parasitic-aware optimization of CMOS RF circuits*. Kluwer Academic Publishers.

[9]   Chan, T., Lu, H., Zeng, J. & Chen, C. (2008). LTCC spiral inductor modeling, synthesis and optimization. *The Asia and South Pacific Design Automation Conference (ASP-DAC)*, 768–771.

[10] Pereira, P., Fino, M. H. & Coito, F.V. (2009). Using discrete-variable optimization for CMOS spiral inductor design. *International Conference on Microelectronics (ICM'09)*, 324–327.

[11] Pereira, P., Fino, M. H., Coito, F. & Ventim-Neves, M. (2009). ADISI- An efficient tool for the automatic design of integrated spiral inductors. *The IEEE International Conference on Electronics, Circuits, and Systems (ICECS'09)*, 799–802.

[12] Pereira P., Fino M.H. ,Coito F. & Ventim-Neves M. (2010). GADISI - Genetic Algorithms Applied to the Automatic Design of Integrated Spiral Inductors. *The Doctoral Conference on Computing, Electrical and Industrial Systems (DoCEIS) - Emerging Trends in Technological Innovation*, Costa de Caparica, Portugal.

[13] Roca, E., Fakhfakh, M., Castro-Lopez, R. & Fernandez, F.V. (2009). Applications of Evolutionary Computation Techniques to Analog, Mixed-Signal and RF Circuit Design – An Overview. *The IEEE International Conference on Electronics, Circuits, and Systems, (ICECS'09)*, 251–254.

[14] S. S. Mohan, M. M. Hershenson, S. P. Boyd & T. H. Lee, Simple Accurate Expressions for Planar Spiral Inductances. *The IEEE Journal of Solid-State Circuits,* vol. 34(10), October 1999.

[15] Aguilera, J., & Berenguer, R. (2003). *Design and Test of Integrated Inductors for RF Applications.* Kluwer Academic Publishers.

[16] Wong, S.-C., Lee, G.-Y., & Ma, D.-J. (2000). Modeling of interconnect capacitance, delay, and crosstalk in VLSI. *IEEE Transactions on Semiconductor Manufacturing ,* vol. 13, 108–111.

[17] Eo, Y., & Eisenstadt, W. (1993). High-speed VLSI interconnect modeling based on S-parameter measurements. *The IEEE Transactions on Components, Hybrids, and Manufacturing Technology,* vol. 16(5), 555–562.

[18] Lai, I. , Fujishima, M. (2008). Design and Modeling of Millimeter-Wave CMOS Circuits for Wireless Transceivers - Era of Sub-100nm Technology, Springer.

[19] Medeiro, F., Rodríguez-Macías, R., Fernández, F.V., Domínguez-Astro, R., Huertas, J.L. & Rodríguez-Vázquez, A. (1994). Global Design of Analog Cells Using Statistical Optimization Techniques. *Analog integrated circuits and signal processing,* vol 6(3), 179–195.

[20] Fakhfakh, M., Cooren, Y., Sallem, A., Loulou, M. & Siarry, P. (2010). Analog Circuit Design Optimization through the Particle Swarm Optimization Technique. *Analog Integrated Circuits & Signal Processing,* vol 63(1), 71–82.

[21] Fakhfakh, M., Sallem, A., Boughariou, M., Bennour, S., Bradai, E., Gaddour, E. & Loulou, M. (2010). Analogue circuit optimization through a hybrid approach. In: *Intelligent Computational Optimization in Engineering: Techniques & Applications,* Springer.

[22] Talbi, E. G. (2002). A taxonomy of hybrid metaheuristics. *Journal of Heuristics, Kluwer Academic Publishers Hingham, MA, USA*, vol 8(5), 541–564.

[23] Siarry, P. & Michalewicz, Z. (2008). *Advances in metaheuristics for hard optimization. Natural Computing Series*, Springer.

[24] Dréo, J., Pétrowski, A., Siarry, P. & Taillard, E. (2006). *Metaheuristics for hard optimization: Methods and case studies.* Springer Verlag.

[25] Daniel, A. (2006). *Evolutionary Computation for Modeling and Optimization.* Springer.

[26] Eiben, A. E & Smith, J. E. (2007). *Introduction to evolutionary computing. Natural Computing Series*, Springer.

[27] Goldberg, D. E. (1989). *Genetic Algorithms in Search, Optimization, and Machine Learning.* Addison-Wesley.

[28] Chan, F. T. S. & Tiwari, M. K. (2007). *Swarm Intelligence: focus on ant and particle swarm optimization.* I-Tech Education and Publishing.

[29] Kennedy, J. & Eberhart, R. C. (1995). Particle swarm optimization. *The IEEE International Conference On Neural Networks*, 1942–1948.

[30] Clerc, M. (2006). *Particle swarm optimization.* International Scientific and Technical Encyclopedia.

[31] Dorigo, M., DiCaro, G. & Gambardella, L. M. (1999). Ant algorithms for discrete optimization. *Artificial life Journal*, 137–172.

[32] Passino, K. M. (2002). Biomimicry of bacterial foraging for distributed optimization and control. *IEEE Control Systems Magazine*, 52-67.

[33] Kirkpatrick, S., Gelatt, C. D. & Vecchi, M. P. (1983). Optimization by Simulated Annealing. *Science, New Series*, vol 220, 671–680.

[34] Glover, F. (1989). Tabu search- part I. *ORSA Journal on Computing.* vol. 1(3), 190–206.

[35] Glover, F. (1990). Tabu search- part II. ORSA Journal on Computing. vol. 2(1), 4–32.

[36] Grimbleby, J. B. (2000). Automatic analogue circuit synthesis using genetic algorithms. *IEE Proceedings-Circuits, Devices and Systems*, vol. 147(6), 319–323.

[37] Mishra, B. K. & Save, S. (2009). Computer Aided Design Automation of Low Power, Low Voltage Four Quadrant Transconductance Multiplier. *The International Conference on Emerging Trends in Engineering and Technology (ICETET)*, 130–134.

[38] Barros, M., Guilherme, J. & Horta, N. (2010). Analog circuits optimization based on evolutionary computation techniques. *Integration, the VLSI Journal*, vol 4 (1).

[39] Reyes-Sierra, M. & Coello-Coello, C. A. (2006). Multi-objective particle swarm optimizers: a survey of the state-of-the-art. *International Journal of Computational Intelligence Research*, vol. (2)3, 287–308.

[40] Fakhfakh, M. (2009). A novel alienor-based heuristic for the optimal design of analog circuits. *Microelectronics Journal*, 141–148.

[41] Haupt, R.L. & Haupt, S.E. (2004). *Practical Genetic Algorithms.* John Wiley & Sons.

[42] Deb, K. (1999). *Multi-Objective Genetic Algorithms: Problem Difficulties and Construction of Test Problems.* Evolutionary Computation, 205–233.

[43] Deb, K., Pratap, A., Agarwal, S. & Meyarivan, T. (2002). A Fast and Elitist Multiobjective Genetic Algorithm: NSGA-II. *IEEE Transactions on Evolutionary Computation*, vol. 6(2), 182–197.

[44] Raquel, C.R. & Naval, P.C. (2005). An Effective Use of Distance in Multiobjective Particle Swarm Optimization. *The Genetic and Evolutionary Computation Conference*, 257–364.

[45] Smith, K.I., Everson, R.M., Fieldsend, J.E., Murphy, C. & Misra, R. (2008). Dominance-Based Multiobjective Simulated Annealing. *The IEEE Transaction on Evolutionary Computation*, vol 12(3), 323–342.

[46] Collette, Y. & Siarry, P. (2003). *Multiobjective optimization principles and case studies*. New York: Springer.

[47] Turkkan, N. (2003). Discrete optimization of structures using a floating-point genetic algorithm. *Annual Conference of the Canadian Society for Civil Engineering*.

[48] *http://tracer-uc3m.es/tws/pso*

[49] *www.swarmintelligence.org*

[50] Li-Pei Wong, Low, M.Y.H. & Chin Soon Chong. (2008). A bee colony optimization algorithm for traveling salesman problem. *The Asia International Conference on Modelling & Simulation*, 818–823.

[51] Passino, K. M. (2002). Biomimicry of bacterial foraging for distributed optimization and control, *IEEE Control Systems Magazine*, 52–67.

[52] Fakhfakh, M., Masmoudi, S., Cooren, Y., Loulou, M. & Siarry, P. (2010). Improving Switched Current Sigma Delta Modulators' Performances Via The Particle Swarm Optimization Technique. *International Journal of Applied Metaheuristic Computing*, vol1(2), IGI-Global, 18–33.

[53] Coello-Coello, C. A. & Lechuga, M. S. (2002). MOPSO: A proposal for multiple objective particle swarm optimization. The evolutionary computation congress. 1051–1056.

[54] Aguilera J., Berenguer R. (2004), *Design and Test of Integrated Inductors for RF Applications*, Kluwer Academic Publishers.

In: Analog Circuits: Applications, Design and Performance
Editor: Esteban Tlelo-Cuautle

ISBN: 978-1-61324-355-8
© 2012 Nova Science Publishers, Inc.

Chapter 9

APPLICATION OF THE ACO TECHNIQUE TO THE OPTIMIZATION OF ANALOG CIRCUIT PERFORMANCES

B. Benhala[1], A. Ahaitouf[1,], M. Kotti[2], M. Fakhfakh[2], B. Benlahbib[1], A. Mecheqrane[1], M. Loulou[2], F. Abdi[1] and E. Abarkane[1]*

[1] University of Sidi Mohamed Ben Abdellah, Faculty of Sciences
and Technology, Fes, Morocco
[2] University of Sfax, Tunisia

Abstract

This chapter presents an adaptation of the Ant Colony Optimization technique to the optimal sizing of analog circuits. Details of the proposed algorithm are given in the following. Performances of the metaheuristic are demonstrated thru test functions. Applications to the optimal sizing of analog circuits, namely, An Operational Amplifier (Op-Amp) and an Operational Transconductance Amplifiers (OTA) are presented. SPICE simulation results are given to show viability of the reached optimal results. A comparison with published works related to the same optimization problems is also given.

Introduction

Nowadays, the realization of very complex integrated electronic circuits and systems is possible thanks to the evolution of Integrated Circuits' technologies. On the other hand, the analog circuit design became so complicated, that the designers generally make appeal to the classical/statistic-based sizing methodologies in order to size their circuits [1]. However, it has been proven that these methods do not guarantee the convergence towards the global optimum solution [2]. 'Mathematic'-based (meta)heuristics were/are also used for this purpose, such as Simulated Annealing (SA) [3], Genetic Algorithms (GA) [4], Tabu Search

---

*E-mail address: ali_ahitouf@yahoo.fr (Corresponding author)

[5,6], etc. However, the efficiency of these methods is highly dependent on the algorithm parameters, the number of variables, the constraint functions and the dimension of the solution space [3]...

In order to overcome drawbacks of the aforementioned optimization approaches, a new set of nature inspired heuristic optimization algorithms was proposed [7]. These techniques are resourceful, efficient and easy to use. They are known as Swarm Intelligence (SI) [8]. They focus on animal behavior and insect conduct in order to develop some metaheuristics which can mimic their problem solution abilities, namely, Particle Swarm Optimization (PSO) [9], Bacterial Foraging Optimization (BFO) [10], Wasp Nets [8], and Ant Colony Optimization (ACO) [11,12].

The Ant System (AS) was proposed in 1996 [12] as a new heuristic for solving optimization problems based on the foraging behavior found in nature. Since its proposal, many variants of AS have been developed since, leading to the formulation of Ant Colony Optimization (ACO) [12-16]. ACO was proposed to solve graph-based problems: the travel salesman problem (TSP) [17,18], vehicle routing problem [19], graph coloring [20],...etc.

In this Chapter we propose an adaptation of the ACO technique to the optimal sizing of analog circuits. According to the knowledge of the authors, it is the first time that ACO is used for such purposes.

The chapter is structured in three main parts: The first part presents an overview of the ACO technique and its adaptation for solving analog optimization problems. The second part shows the viability of the proposed algorithm via some test functions. The third part deals with the application of the proposed algorithm to the optimal design of analog circuits. Two sizing/optimization problems are showcased; namely the optimal sizing of an operational amplifier (Op-Amp) and an operational transconductance amplifier (OTA). SPICE simulation results are provided to show the viability of the proposed algorithm. Comparisons with published works related to the same optimization problems are also given and discussed.

Ant Colony Optimization

I. Ant Colony Optimization Technique: An Overview

ACO technique is inspired by the collective behavior of deposit and monitoring of slopes that is observed in insect colonies [13], such as ants. Figure 1 shows an illustration of the ability of ants to find the shortest path between food and their nest. It is illustrated through the example of the appearance of an obstacle on their path. Ants communicate indirectly through dynamic changes in their environment (pheromone trails).

Initially, ACO technique was proposed to deal with graph problems. A graph is composed of vertices and edges. Each ant constructs its own path from the starting vertex to the final one by 'walking' along edges connecting the vertices while deposing a certain amount of pheromones (a chemical substance) that evaporates during the time, unless it is reinforced by another ant 'walking' along the same edge. Thus, the 'best', i.e. the shortest, path is determined on the base of these pheromones. Besides, movement of the ants is highly conditioned by their visibility regarding the final objective.

Application of the ACO Technique to the Optimization ...

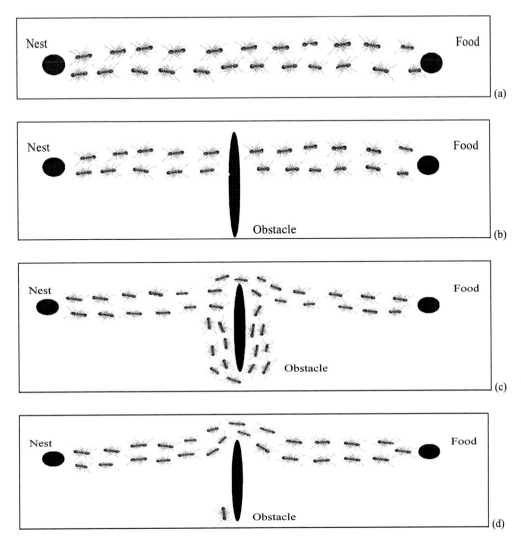

Figure 1. Self-adaptive behavior of a real ant colony. (a) Ants follow a path between nest and food source; (b) An obstacle appears on the path: ants choose, with equal probability, whether to turn left or right; (c) Higher amount of pheromone is deposited on the short path; (d) The majority of ants have chosen the shortest path.

Practically, for solving such problems, ants randomly select the vertex to be visited. When ant k is in vertex i, the probability P_{ij}^{k} of going to vertex j is given by expression (1) [12,13,16].

$$
p_{ij}^{k} = \begin{cases} \dfrac{(\tau_{ij})^{\alpha} \cdot (\eta_{ij})^{\beta}}{\sum\limits_{l \in J_i^k} (\tau_{il})^{\alpha} \cdot (\eta_{ij})^{\beta}} & \text{if } j \in J_i^k \\[2em] 0 & \text{if } j \notin J_i^k \end{cases} \tag{1}
$$

where J_i^k is the set of neighbors of vertex i of the kth ant, τ_{ij} is the amount of pheromone trail on edge (i, j), α and β are weightings that control the pheromone trail and the visibility value, i.e η_{ij}, which expression is given by (2).

$$
\eta_{ij} = \frac{1}{d_{ij}} \tag{2}
$$

d_{ij} is the distance between vertices i and j.

Pheromone values are updated at each iteration by all the m ants that have built a solution in the iteration itself. The pheromone τ_{ij}, which is associated with the edge joining vertices i and j, is updated as follows:

$$
\tau_{ij} = (1 - \rho) \cdot \tau_{ij} + \sum\nolimits_{k=1}^{m} \Delta \tau_{ij}^k \tag{3}
$$

where ρ is the evaporation rate, $\Delta \tau_{ij}^k$ is the quantity of pheromone laid on edge (i, j) by ant k:

$$
\Delta \tau_{ij}^k = \begin{cases} \frac{Q}{L^k} & \text{if ant } k \text{ used edge } (i, j) \text{ in its tour} \\[1em] 0 & \text{otherwise} \end{cases} \tag{4}
$$

Q is a tunable parameter and L^k is the length of the path constructed by ant k.

II. Adaptation of the ACO Technique

In order to deal with analog sizing optimization problems, we propose an adaptation of the basic ACO technique to such applications. The idea consists of building an equivalent graph that can represent the optimization variables as well as the objective functions. The vertices of the graph correspond to the discrete variable values. Thus, each ant will construct its path by a (random) move from a variable value to another, as it is shown in Figure 2, where $V1$, $V2$, $V3...VN$ are the discrete variable vectors [21].

In short, each ant k will randomly choose a path (values of $V1$, $V2$ ...), according to the probability given by expression (1), and form a directed graph while randomly generating a rate of pheromone at the formed graph edges. At each iteration the path giving the minimum value of the objective function (OF) sees its rate increases, in contrast with the other paths which pheromone rates are partially evaporated with respect to expression (3).

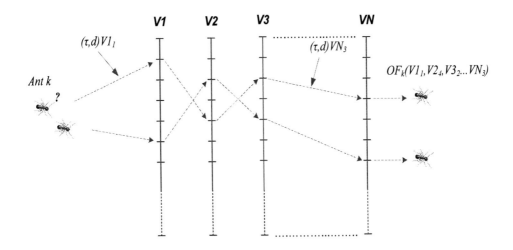

Figure 2. A pictorial graph showing the mouvement of the ants.

Regarding the visibility parameter, τ_{ij} is considered to be the current absolute value of the difference between the best so far memorized value and the current value of OF_k (the kth OF).

The proposed algorithm operates with respect to the following three routines:

1. REFERENCE point ⎫
2. Calculate DISTANCE ⎬ Initialization
3. 'ACO' algorithm

These three routines are summarized below.

- REFERENCES: it is an initialization routine: its main aim consists of finding a 'good' starting point, which is equivalent to the determination of a 'good' initial reference for computing the *OF* relative distance. Thus, the distance heuristic is shunted in this first phase, i.e. $\beta=0$. Figure 3 shows the initialization routine's flowchart.
- DISTANCE: it consists of the computation of the distance value (*dV*) for each discrete variable. Actually, this phase consists of an update of the 'optimal' *OF* (*OFop(t)*) found up to the current iteration (*t*). The corresponding flowchart is presented in Figure 4. *OFop(t)* is considered as the reference for computing the distance d_{ij} at iteration (*t+1*). Expression (5) gives the value of *dV1i*,

$$dV1_i = OF(V1_i, V2_{ref}, V3_{ref}, ...) - OF_{op} \tag{5}$$

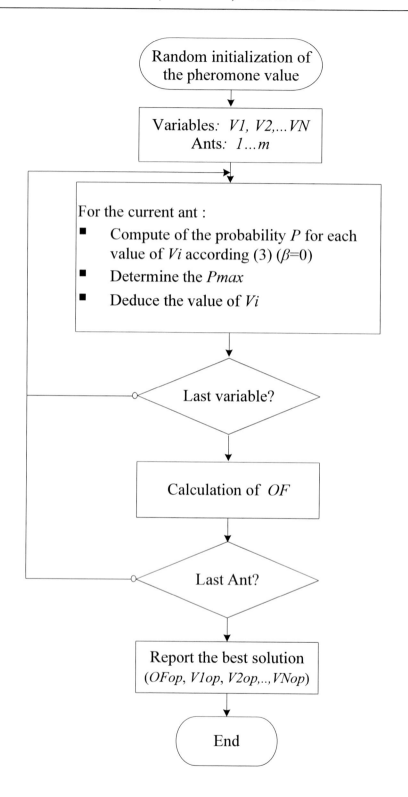

Figure 3. Flowchart "REFERENCE".

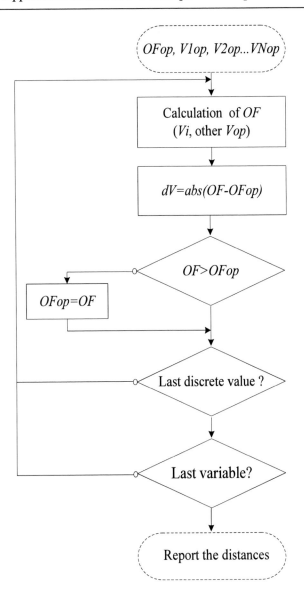

Figure 4. Flowchart "DISTANCES".

- 'ACO': it is the core of the ACO program. It is shown in Figure 5: it mainly consists of including computing the movement probability of each ant, calculation and update of the pheromone rates and taking into consideration the update of the distance references (i.e. visibility values).

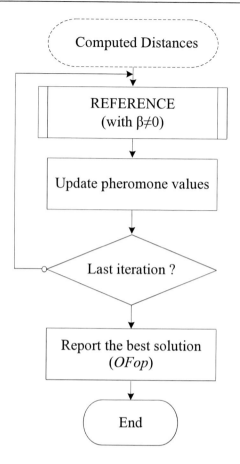

Figure 5. Flowchart "ACO".

Tests Function Examples

In order to check performances of the proposed algorithm, two test functions [22,23] were used for this purpose. In Figures 6 and 7 we present a plot of these two functions which expressions are given in Table I.

Table I. The test functions

	Variable Bounds	Objective Functions
F1	$-4 \leq x, y \leq 4$	$f(x,y) = (1.5 - x + xy)^2 + (2.25 - x + xy^2)^2 + (2.625 - x + xy^3)^2$
F2	$0 \leq x, y \leq 10$	$f(x,y) = x\sin(4x) + 1.1y\sin(2y)$

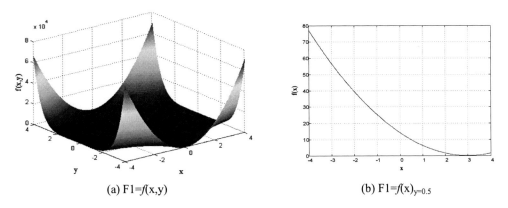

(a) F1=$f(x,y)$ (b) F1=$f(x)_{y=0.5}$

Figure 6. Test function F1.

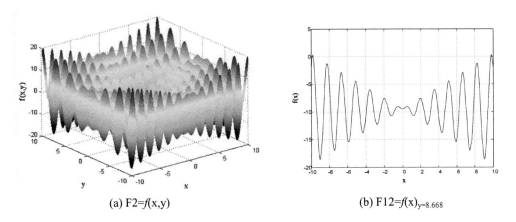

(a) F2=$f(x,y)$ (b) F12=$f(x)_{y=8.668}$

Figure 7. Test function F2.

Table II gives ACO algorithm parameters.

Table II. Parameters of ACO algorithm

Number of iterations	10
Number of Ants	10
Evaporation rate: ρ	0.1
Quantity of deposit pheromone by the best ant: Q	0.2
Pheromone factor: α	1
Heuristics factor: β	1

Table III presents a comparison between ACO algorithm results and the theoretical ones.

Table III. Comparison of the ACO algorithm results and the theoretical ones

	Theoretical results		ACO algorithm results	
	Parameters	Objective function	Parameters	Objective function
F1	(3,0.5)	0	(3,0.5)	0
F2	(9.039,8.668)	-18.5547	(9.039,8.667)	-18.5546

Good agreement between theoretical results and those obtained using the proposed MATLAB-implemented ACO algorithm can be noticed from results shown in Table III.

Application to the Optimal Sizing of Analog Circuits

In the following we present the application of the proposed adaptation of the ACO technique to the optimal sizing of analog circuits. Two analog CMOS circuits are considered: an operational amplifier (Op-Amp) and an operational transconductance amplifier (OTA). In this section, we give optimization results and present comparison with results obtained using the geometric programming technique in the case of the Op-Amp [24], and using an algorithm driven heuristic, which was proposed in [25,26], in the case of the OTA. SPICE simulations are given to show viability of obtained results.

I. Example 1: An Operational Amplifier

The schematic of two-stage CMOS Op-Amp is shown in Figure 8.

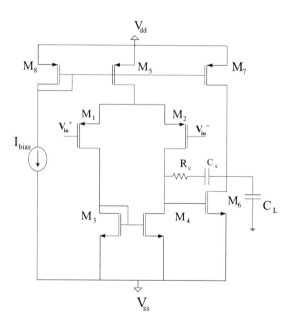

Figure 8. A two stage CMOS operational amplifier.

In short, the circuit consists of an input differential stage with an active load followed by a common-source stage also with active load. This amplifier is assumed to be part of a very large scale integration (VLSI) system and is only required to drive a fixed on-chip capacitive load of a few picofarads. This Op-Amp architecture has many advantages: high open-loop voltage gain, rail-to-rail output swing, large common-mode input range, only one frequency compensation capacitor, and a small number of transistors.

The design of Op-Amp continues to pose a challenge as transistor channel lengths scale down with each generation of CMOS technologies [27].

OBJECTIVE FUNCTIONS AND CONSTRAINTS

Performances of an Op-Amp are evaluated via several parameters such as:

- Open-loop voltage gain Av:

$$Av = \frac{2C_{ox}}{(\lambda_n + \lambda_P)^2} \sqrt{\frac{\mu_n \mu_P}{I_1 I_7} \left(\frac{W}{L}\right)_2 \left(\frac{W}{L}\right)_6} \tag{6}$$

- Power dissipation P:

$$P = (V_{dd} - V_{ss})(I_{bias} + I_5 + I_7) \tag{7}$$

V_{dd} and V_{ss} are respectively the positive and the negative supply voltages,

- Common mode rejection ratio $CMRR$:

$$CMRR = \frac{2C_{ox}}{(\lambda_n + \lambda_P)\lambda_P} \sqrt{\frac{\mu_n \mu_P}{I_5^2} \left(\frac{W}{L}\right)_1 \left(\frac{W}{L}\right)_3} \tag{8}$$

- The unity-gain bandwidth (ω_c) can be approximated as the open-loop gain times the 3-dB bandwidth:

$$\omega_c \approx \frac{gm_1}{Cc} \tag{9}$$

- Die area A:

$$A \approx \sum_{i=1}^{8} W_i L_i \tag{10}$$

$W_1 ... W_8$ and $L_1 ... L_8$ are widths and lenghs of transistors $M_1 ... M_8$ respectively. I_{bias} is the current bias, gm refers to the transconductance of the MOS transistor, C_{ox}, λ_n, λ_p, μ_n and μ_p are technological parameters. Cc is a compensation capacitor.

Determining the optimal sizing of the transistors for a specific design involves a tradeoff among all these performance measures.

Regarding imposed constraints; all the transistors have to operate in the saturation mode. Expressions (11)-(14) give the corresponding constraints that have to be satisfied when computing optimal sizes of the transistors M_1 (and M_2), M_5, M_6 and M_7 respectively.

$$V_{cm,min} - V_{ss} - V_{TP} - V_{TN} \geq \sqrt{\frac{2I_1}{\mu_n C_{ox}\left(\dfrac{W}{L}\right)_3}} \tag{11}$$

$$V_{dd} - V_{cm,max} + V_{TP} \geq \sqrt{\frac{2I_1}{\mu_P C_{ox}\left(\dfrac{W}{L}\right)_1}} + \sqrt{\frac{2I_5}{\mu_P C_{ox}\left(\dfrac{W}{L}\right)_5}} \tag{12}$$

$$V_{out,min} - V_{ss} \geq \sqrt{\frac{2I_7}{\mu_n C_{ox}\left(\dfrac{W}{L}\right)_6}} \tag{13}$$

$$V_{dd} - V_{out,max} \geq \sqrt{\frac{2I_7}{\mu_P C_{ox}\left(\dfrac{W}{L}\right)_7}} \tag{14}$$

where $I_5 = \dfrac{\left(\dfrac{W}{L}\right)_5}{\left(\dfrac{W}{L}\right)_8} I_{bias}$, $I_7 = \dfrac{\left(\dfrac{W}{L}\right)_7}{\left(\dfrac{W}{L}\right)_8} I_{bias}$, and $I_1 = \dfrac{I_5}{2}$ while respecting expression (16).

I_i is the drain current of transistor i, and μCox is a technology parameter. V_{TP} and V_{TN} are the PMOS and the NMOS threshold voltage, respectively.

$$\frac{\left(\dfrac{W}{L}\right)_3}{\left(\dfrac{W}{L}\right)_6} = \frac{\left(\dfrac{W}{L}\right)_4}{\left(\dfrac{W}{L}\right)_6} = \frac{1}{2}\frac{\left(\dfrac{W}{L}\right)_5}{\left(\dfrac{W}{L}\right)_7} \tag{15}$$

OPTIMIZATION RESULTS

Actually, the considered optimization problem is a multi-objective one that consists of maximizing the open loop voltage gain and the common mode rejection ratio, and minimizing both the die area and the consumed power.

In the literature, there are mainly two approaches that are proposed to deal with multi-objective problems. i.e. the weighted cost functions, and the Pareto tradeoff front [22]. In this work we focus on the weighted cost functions technique. It consists of pondering each function and transforming the set of objectives into an equivalent cost function (OF_{eq}). Thus, OF_{eq} can be expressed as follows:

$$OF_{eq} = -\alpha_1 A_v + \alpha_2 P + \alpha_3 A - \alpha_4 CMRR - \alpha_5 \omega_c \qquad (16)$$

where α_{1-5} are normalization coefficients.

The proposed ACO algorithm was used to optimize OF_{eq}. The algorithm's parameters are given in Table IV:

Table IV. Parameters of the ACO algorithm

Number of iterations	50
Number of ants: m	20
Evaporation rate: ρ	0.1
Quantity of deposit pheromone by the best ant: Q	0.2
Pheromone factor: α	1
Heuristics factor: β	1

Table V presents a comparison between results obtained using the proposed ACO algorithm and those proposed in [24] that were reached using the geometric programming optimization technique. One can easily notice that ACO performances are the 'best'.

Table V. Optimal device sizing and performances

Specifications	ACO Optimization	[24]
Technology	AMS 0.35 µm	AMS 0.8 µm
Voltage supply [V]	+5/0	+5/0
(W[µm]/L[µm])$_{1,2}$	201.00/0.80	232.80/0.80
(W[µm]/L[µm])$_{3,4}$	130.40/0.80	143.60/0.80
(W[µm]/L[µm])$_5$	51.20/0.80	64.60/0.80
(W[µm]/L[µm])$_6$	570.40/0.80	588.80/0.80
(W[µm]/L[µm])$_7$	111.70/0.80	132.60/0.80
(W[µm]/L[µm])$_8$	5.00/0.80	2.00/0.80
Av[dB]	102.1	89.2
$CMRR$[dB]	106.0	92.5
P[mW]	1.7	5.0
A[µm²]	1124	8200
ω_c [MHz]	98	86

SPICE simulation results performed, using AMS 0.35μm technology, are presented in the following figures; they show the good agreement with the expected ones. Corresponding performances are summarized in Table VI.

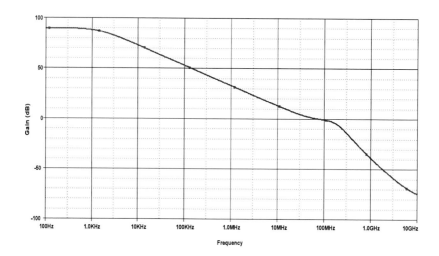

Figure 9. SPICE simulation results for voltage gain.

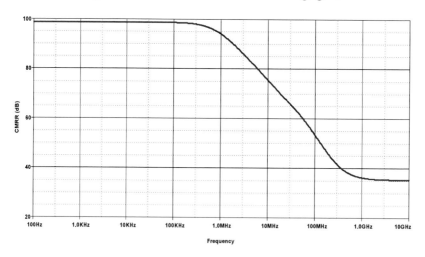

Figure 10. SPICE simulation results for the common mode rejection ratio.

Table VI. Reached Performances

Performances	Theoretical (MATLAB)	Simulation (SPICE)
Av [dB]	102.1	90.0
$CMRR$ [dB]	106.00	98.66
ω_c [MHz]	98	102
P [mW]	1.7	2.1

II. Example 2: A CMOS Folded Cascode OTA

An OTA is basically an Op-Amp without any output buffer, preventing it from driving resistive or large capacitive loads. OTAs are preferred over Op-Amps mainly because of their smaller size and their simplicity. The transconductance Op-Amp (OTA) differs from the standard Op-Amp by a high gain, a minimal current polarization and a large bandwidth.

Different OTA architectures are available in the literature. In the following we deal with optimizing the sizing of a fully differential folded cascade OTA [26,29].

Figure 11 shows a CMOS folded cascode operational transconductance amplifier which has a differential stage consisting of NMOS transistors M_9 and M_{10}. Transistors M_{11} and M_{12} provide the DC bias voltages to (M_1,M_2) and (M_3,M_4) transistors, while cascode transistors (M_5-M_8) are controlled respectively by transistors M_{13} and M_{14}.

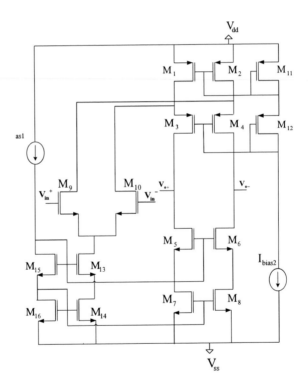

Figure 11. A folded cascode OTA.

OBJECTIVES FUNCTIONS AND CONSTRAINTS

Expressions (17)-(20) give expressions of the main performances of the OTA, i.e. the open-loop voltage gain (Av), the unity-gain frequency (Ft), the power-supply rejection ratio ($PSRR$) and the common mode rejection ratio ($CMRR$) [28]:

$$Av = g_{m9} R_{out} \qquad (17)$$

$$Ft = \frac{g_{m9}}{2\pi C_L} \tag{18}$$

$$PSRR = \frac{(R_1 + R_2 + r_3)(R_1 + r_3 + g_{m3}r_3R_1)}{R_2(1 + g_{m3}r_3)} \tag{19}$$

$$CMRR = \frac{R_{out}(1 + 2g_{m9}(r_{13} + r_{14}))}{2(r_3 + R_1)} \tag{20}$$

where: $R_{out} = R_2//(g_{m3}r_3R_1)$, $R_2 = g_{m5}r_5r_{07}$ and $R_1 = r_1//r_9$.

g_{m3}, g_{m5} and g_{m9} are respectively the transconductances of transistors M3, M5 and M9. r_1, r_3, r_5, r_7, r_9, r_{13} and r_{14} are respectively the drain to source resistances of transistors M1, M3, M5, M7, M9, M13 and M14, and C_L is the load capacitance.

Expressions (21)-(24) give expressions of the constraints to be satisfied to insure saturation of the MOS transistors encompassing the OTA.

$$Vdd - V_{out,\max} \geq \sqrt{\frac{3I1}{Kp\left(\dfrac{W}{L}\right)_1}} + \sqrt{\frac{2I1}{Kp\left(\dfrac{W}{L}\right)_3}} \tag{21}$$

$$V_{out,\min} - Vss \geq \sqrt{\frac{I1}{Kn\left(\dfrac{W}{L}\right)_5}} + \sqrt{\frac{I1}{Kn\left(\dfrac{W}{L}\right)_7}} \tag{22}$$

$$Vdd - V_{in,\max} + V_{tn} \geq \sqrt{\frac{3I1}{Kp\left(\dfrac{W}{L}\right)_1}} \tag{23}$$

$$V_{in,\min} - Vss - V_{tn} \geq \sqrt{\frac{I1}{Kn\left(\dfrac{W}{L}\right)_9}} + \sqrt{\frac{2I1}{Kn\left(\dfrac{W}{L}\right)_{13}}} + \sqrt{\frac{2I1}{Kn\left(\dfrac{W}{L}\right)_{14}}} \tag{24}$$

$V_{in,min}$, $V_{in,max}$, $V_{out,min}$, $V_{out,max}$ are respectively the minimum and the maximum allowed input and output voltages. Kn and Kp are technology parameters. I_i is the drain current of transistor i.

OPTIMIZATION RESULTS

The weighting technique was also used to transform the multi-objective problem into a mono-objective one. The equivalent OF expression is given by expression (25).

$$OF = \alpha_1 A_v + \alpha_2 F_t + \alpha_3 PSRR + \alpha_4 CMRR \tag{25}$$

α_{1-4} are normalization coefficients.

The static and dynamic performances of the OTA are set according to the application specifications. Table VII presents specifications taken from [25] and that correspond to a Wide band CMFB Switched capacitor application. The technology under consideration is AMS 0.35μA.

Table VII. Specifications

Av [dB]	≥ 60
GBW[MHz]	≥ 200
C_L[pF]	0.1
Vdd[V]/Vss[V]	+1.8/-1.8
I_{bias1}[μA]	60
I_{bias2}[μA]	90

The proposed ACO algorithm was used to compute the optimal values of the geometric dimensions of the MOS transistors forming the OTA. Obtained sizings are summarized in Table VIII. In addition, a comparison with performances and sizings proposed in [26,29] which were reached using an optimization heuristic, is given. One can clearly notie that the ACO results are the 'best' ones.

Table VIII. Optimal device sizing and performances

Specifications	Proposed Algorithm	[26]
Technology	AMS 0.35 μm	
Voltage supply [V]	±1.8	
$(W$[μm]/L[μm]$)_1$	46.82/1.00	34.85/1.00
$(W$[μm]/L[μm]$)_3$	30.91/1.00	23.00/1.00
$(W$[μm]/L[μm]$)_5$	50.00/1.00	47.15/1.00
$(W$[μm]/L[μm]$)_9$	50.00/1.00	49.90/1.00
$DC\ Gain$ [dB]	84.15	82.89
GBW [MHz]	534.00	533.55
$CMRR$ [dB]	94.82	93.56
$PSRR$ [dB]	84.57	73.23

Figures 12-14 show SPICE simulation results of the gain, the common mode rejection ratio and the power supply rejection ratio, respectively.

These results are in good agreement with the expected theoretical ones. Table IX summarizes and compares reached performances.

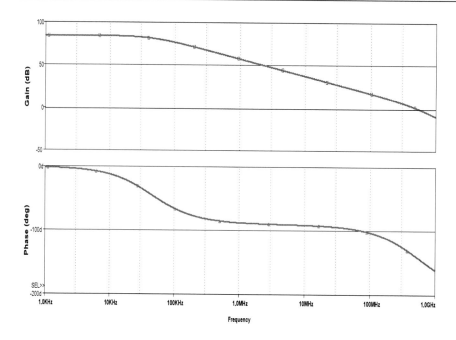

Figure 12. SPICE simulation results of gain and phase.

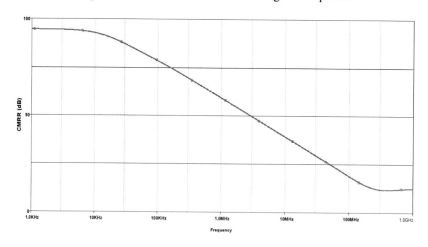

Figure 13. SPICE simulation results of the common mode rejection ratio.

Table IX. Reached performances

Performances	Theoretical (MATLAB)	SPICE Simulations
Av [dB]	84.15	84.08
Ft [MHz]	534.00	507.51
$CMRR$ [dB]	94.82	94.52
$PSRR$ [dB]	84.15	79.57

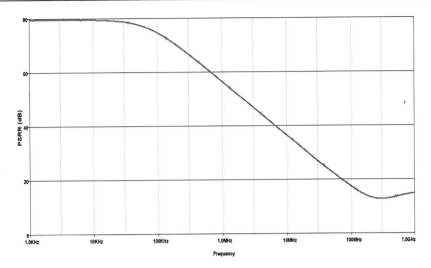

Figure 14. SPICE simulation results for the power-supply rejection ratio (PSRR).

Conclusion

The presented work proposes a novel adaptation of the ant colony optimization technique for dealing with the optimal sizing of analog circuits. The proposed algorithm was first validated thru mathematical test functions and then by application to the optimal sizing of two CMOS analog circuits, namely an operational amplifier and an operational transconductance amplifiers. Obtained sizing and reached results were compared to published works. The proposed ACO algorithm offers better results in terms of objectives. Viability of the technique was proved via SPICE simulations.

References

[1] Medeiro, F., Rodríguez-Macías, R., Fernández, Domínguez-Astro, F.V.R., Huertas, J.L. & Rodríguez-Vázquez, A. (1994). Global design of analog cells using statistical optimization techniques. *Analog integrated circuits and signal processing,* **6**(3), 179-195.
[2] Talbi, E. G. (2002). A taxonomy of hybrid metaheuristics. *Journal of Heuristics,* **8**(2), 541-564.
[3] Dreo, J., Petrowski, A., Siarry, P. & Taillard, E. (2006). Metaheuristics for hard optimization: methods and case studies. *Springer.*
[4] Grimbleby, J. B. (2000). Automatic analogue circuit synthesis using genetic algorithms. IEE Proceedings-Circuits, Devices and Systems, **147**(6), 319-323.
[5] Glover, F. (1989). Tabu search-part I. *ORSA Journal on computing,* **1**(3), 190-206.
[6] Glover, F. (1990). Tabu search-part II. *ORSA Journal on computing,* **2**(1), 4-32.
[7] Roca, E., Fakhfakh, M., Castro-López, R. & Fernández, F. V. (2009). Applications of evolutionary computation techniques to analog, mixed-signal and RF circuit design. *The*

IEEE international conference on electronics, circuits, and systems (ICECS). Hammamet, Tunisia.

[8] Chan, F. T. S. & Tiwari, M. K. (2007). Swarm Intelligence: focus on ant and particle swarm optimization. *I-Tech Education and Publishing*.

[9] Fakhfakh, M., Cooren, Y., Sallem, A., Loulou, M. & Siarry, P. (2010). Analog circuit design optimization through the particle swarm optimization technique. *Journal of Analog Integrated Circuits & Signal Processing.* **63**(1), *Springer* , 71-82.

[10] Chatterjee, A., Fakhfakh, M. & Siarry, P. (2010). Design of second generation current conveyors employing bacterial foraging optimization. *Microelectronics Journal,* **41**(10), *Elsevier*, 616-626.

[11] Dorigo, M. & Stüzle, T. (2004). Ant Colony Optimization. *MIT Press*.

[12] Marco, D., Mauro, B. & Thomas, S. (2006). Ant colony optimization. *IEEE computational intelligence magazine,* **11**. pp. 28-39.

[13] Dorigo, M., Di-Caro, G. & Gambardella, L. M. (1999). Ant algorithms for discrete optimization. *Artificial Life Journal,* **5**, 137-172.

[14] Shtovba, S. D. (2005). Ant algorithms: Theory and applications. *Programming and Computer Software,* **31** (4), 167-178.

[15] Dorigo, M., Maniezzo, V. & Colorni, A. (1996). The ant system: Optimization by a colony of coorperating agents. *IEEE Transactions on Systems, Man and Cybernetics-Part B,* **26**(1), 29-41.

[16] Stutzle, T. & Hoos, H.H. (2000). MAX–MIN Ant System. *Future Generation Computer Systems,* **16**(8), 889-914.

[17] Dorigo, M. & Gambardella, L. M. (1997). Ant colonies for the traveling salesman problem. *BioSystems,* **43**, 73-81.

[18] Jinhui, Y., Xiaohu, S., Maurizio, M. & Yanchun, L. (2008). An ant colony optimization method for generalized TSP problem. *Progress in Natural Science,* **18**(11), 1417-1422.

[19] Yu, B., Yang, Z. & Yao, B. (2009). An improved ant colony optimization for vehicle routing proble. *European Journal of Operational Research,* **196**, 171-176.

[20] Salari, E. & Eshghi, K. (2005). An ACO algorithm for graph coloring problem. *Congress on Computational Intelligence Methods and Applications (CIMA), Turkey.*

[21] Benhala, B., Ahaitouf A., Mechaqrane A., Benlahbib B., Abdi F., Abarkan E.& Fakhfakh M. (2011). Sizing of current conveyors by means of an ant colony optimization technique. *The International Conference on Multimedia Computing and Systems (ICMCS'11), Morocco.*

[22] Haupt, R. L. & Haupt, S. E. (2004). *Pratical Genetic Algotithms*. Willey.

[23] *www-optima.amp.i.kyoto-u.ac.jp/member/student/hedar/Hedar_files/TestGO_files*

[24] Hershenson, M. M., Boyd, S. P. & Lee, T. H. (2001). Optimal Design of a CMOS Op-Amp via Geometric Programming. *The IEEE Transactions on Computer-Aided Design of Integrated Circuits and Systems,* **20**(1). pp. 1-21.

[25] Zhang, L. (2005). System and circuit design techniques for WLAN-enabled multi-standard receiver. *PhD Dissertation, The Ohio State University.*

[26] Daoud, H., Bennour, S., Salem, S. B. & Loulou, M. (2008). Low power SC CMFB folded cascode OTA optimization. *The IEEE International Conference on Electronics, Circuits and Systems (ICECS), Malta.*

[27] Razavi, B. (2002). *Design of Analog CMOS Integrated Circuits.* McGraw-Hill International Edition: Electrical Engineering Series.

[28] Stefanovic, D. & Kayal, M. (2008). *Structured analog CMOS design*. Springer.

[29] Daoud, H., Ben Salem, S., Zouari, S. & Loulou, M. (2006). Design of fast OTAs in different MOS operating modes using 0.35μm CMOS process. *IEEE International Conference on microelectronics (ICM)*, Pakistan.

In: Analog Circuits: Applications, Design and Performance
Editor: Esteban Tlelo-Cuautle

ISBN 978-1-61324-355-8
© 2012 Nova Science Publishers, Inc.

Chapter 10

ANALOG MISMATCH ANALYSIS BY STOCHASTIC NONLINEAR MACROMODELING

Xue-Xin Liu[1], Yao Yu[2], Hai Wang[1] and Sheldon X.-D. Tan[1]

[1]Department of Electrical Engineering, University of California,
Riverside, CA 92521
[2]School of EEE, Nanyang Technological University,
Singapore 639798

Abstract

This chapter presents a fast non-Monte-Carlo method to calculate mismatch in time domain for analog and mixed-signal circuits in the nanometer regime. The new method describes the local random mismatch by a noise source with an explicit dependence on geometric parameters, and is further expanded by stochastic orthogonal polynomials. This forms a stochastic differential-algebra-equation. To deal with large-scale problems, the stochastic differential-algebra-equation is linearized at a number of snapshots along the nominal transient trajectory, and hence is naturally embedded into a trajectory-piecewise-linear (TPWL) macromodeling. We further show the improvement of the PWL by a novel incremental aggregation of subspaces identif ed at those snapshots. Numerical results show that the presented method, *isTPWL*, is hundreds of times faster than Monte-Carlo method with a similar accuracy. On top of this, the new macromodel further reduces runtime signif cantly.

Keywords: Mismatch, analog circuits, stochastic, nonlinear modeling.

1. Introduction

Transistor mismatch is the primary obstacle to reach a high-yield rate for analog designs in sub-90nm technologies. For example, due to an inverse-square-root-law dependence with the transistor area, the mismatch of CMOS devices nearly doubles for every process generation less than 90nm [5, 6]. Since the traditional worst-case or corner-case based analysis is too pessimistic to sacrif ce the speed, power, and area, the statistical approach [1,5,7,11,12] thereby becomes a viable approach to estimate the analog mismatch. Same as the process variation, there are two types of mismatch. One is systematic (or global spatial variation) and the other is stochastic (or local random variation). This chapter focuses on the stochastic one. Analog circuit designers usually perform a Monte-Carlo (MC) analysis to analyze the stochastic mismatch and predict the statistical functionality of their designs. As MC analysis requires a large number of repeated circuit simulations, its computational cost is expensive. Moreover, the pseudo-random generator in MC introduces numerical noises that may lead to errors.

There are many non-Monte-Carlo methods [1,5,11] developed recently for the stochastic mismatch analysis. [11] first calculated dc sensitivities with respect to small device-parameter perturbations and scaled them as desired mismatches. [5] extended [11] by modeling dc mismatches as ac noise sources. The mismatch, defined in a transient simulation, is converted back from the power spectral density (PSD) in frequency-domain. The speed of these equivalent mismatch simulations is hundred times faster than the Monte-Carlo approaches but the accuracy remains a concern.

Recently, SiSMA [1] studied the mismatch within the framework of the stochastic differential-algebra-equation (SDAE) similar to deal with the transient noise [3]. Due to the introduction of the random variable into the DAE, it is unknown if the derivative is still continuous. Moreover, designers' top interest is the mismatch of the channel current in CMOS transistors. SiSMA thereby modeled the mismatch as a stochastic current-source and formed a SDAE. As such, by assuming that the magnitude of the stochastic mismatch is much smaller than the nominal case, SiSMA linearized the nominal SDAE at dc with the stochastic current source. The obtained dc solution is used as an initial condition (ic) for the transient analysis. As the stochastic current source is only included during dc, this assumption may not hold to accurately describe the mismatch during the transient simulation. Moreover, to avoid an expensive Monte-Carlo simulation, SiSMA calculated the mismatch by the extraction and analysis of a covariance matrix. It would be slow to analyze the covariance matrix for thousands of devices. In addition, the entire circuit is analyzed twice and is computationally expensive for large-scale problems. Therefore, there is still a need to develop a fast transient mismatch analysis that requires improvements in two-fold: a different non-Monte-Carlo method and an efficient macromodel by the nonlinear model order reduction.

This chapter presents a fast non-Monte-Carlo mismatch analysis by an incremental and stochastic trajectory piecewise linear macromodel, namely *isTPWL method*. First, we introduce a transient mismatch model and its macromodeling. Then we show how to linearize the SDAE along a number of snapshots on a nominal transient trajectory, and add the stochastic current source (for mismatch) at each snapshot as a perturbation. This is more accurate than considering the mismatch through an ic condition [1]. Along the snapshots of the nominal transient trajectory, we further show how to apply an improved trajectory-piecewise-linear (TPWL) model order reduction [2,13,15] to generate a stochastic nonlinear macromodel, and apply it for a fast transient mismatch analysis along the full transient trajectory. The presented approach applies an incremental aggregation on those local tangent subspaces, linearized at snapshots. It reduces the computational complexity of [2] yet improves the accuracy of [13]. As shown by experiments, our TPWL method is 5 times more accurate than the work in [13] and is 20 times faster than the work in [2] on average. In addition, utilizing nonlinear macromodels reduces the runtime by up to 25 times compared to the use of the full model during the mismatch analysis.

Next, in order to efficiently solve the SDAE without applying the Monte-Carlo iterations or analyzing the expensive co-variance matrix [1], we show how to describe the stochastic variation by a spectral stochastic method based on stochastic orthogonal polynomials (SOPs) and form an according SDAE [17]. The SOPs have been applied to deal with the linear interconnection variation in [16] and power grid analysis [8, 9]. The chapter present a new method to apply SOPs for nonlinear analog circuits during a non-Monte-Carlo mismatch analysis. Experiments show that compared to the Monte-Carlo method, our method is 1000 times faster with a similar accuracy.

The rest of the chapter is organized in the following manner. In Section 2., we present the background of the mismatch model and the nonlinear model order reduction. In Section 3., we discuss a transient mismatch analysis in SDAE, including a perturbation analysis and a non-Monte-Carlo analysis by the SOP expansions. We develop an incremental and stochastic TPWL model order reduction for mismatch in Section 4.. We present numerical results in Section 5., and conclude the chapter in Section 6..

2. Preliminary

2.1. Mismatch Model

The mismatch model and analysis is the key to a precision analog circuit design such as ADC/DACs. There are global mismatch and local mismatch. The local mismatch is the most difficult one to analyze and hence it is the focus of this chapter.

The local mismatch is stochastic and process-parameter dependent. Most CMOS mismatch models are based on the Pelgrom's work [12] that relates the local mismatch variance of one electrical parameter (such as the channel current I_d) with geometrical parameters (such as the area A) by a geometrical dependence equation

$$\sigma_{I_d} = \frac{\kappa^\beta}{\sqrt{A}} \tag{1}$$

for two devices closely laid out, i.e. a local variation. Note that $A = W \cdot L$ is the area of a width W and length L, and κ^β is an extracted constant depending on the operating region β.

For other transistors such as diode, BJT and etc. and to consider process parameters other than the geometry, a more general purposed mismatch model can be derived through a so-called backward propagation of variance (BPV) method [7]. For example, the base-current I_b depends on the emitter area, sheet resistance and base current density. The BPV model then relates the local mismatch of an electrical property e with those process parameters p_l by a first-order sensitivity equation

$$\sigma_e = \sum_l (\frac{\partial e}{\partial p_l})\sigma_{p_l}. \tag{2}$$

Based on the above mismatch model, we introduce a non-Monte-Carlo transient mismatch analysis for a large number of transistors in Section 3.

2.2. Nonlinear Model Order Reduction

The nominal nonlinear circuit is described by the following differential-algebra-equation (DAE)

$$f(x,\dot{x},t) = \mathcal{B}u(t),\qquad(3)$$

where x ($\dot{x} = dx/dt$) is the state variable including nodal voltage and branch current, $f(x,\dot{x},t)$ is to describe the nonlinear $i-v$ relation, and $u(t)$ is the external sources with a topology matrix $\mathcal{B}$ describing how to add them into the circuit. The time to solve (3) comes from three parts: device evaluation, matrix factorization, and time-step control and integration. When the circuit size is large or when devices are latent in most of time, the portion of runtime mainly comes from the matrix factorization. Under this condition, the use of model order reduction to reduce the circuit size is effective to reduce the overall runtime and hence can be applied in a transient mismatch analysis as well.

Model order reduction is basically to f nd a small dimensioned subspace that can represent the original state space with a preserved system response. This can be usually realized in the view of a coordinate transformation. For linear circuits, the coordinate transformation can be described by a linear mapping

$$z = V^T x, \quad x = Vz,$$

where a small dimensioned projection matrix V ($\in N \times q$, $q << N$) can be constructed from the f rst few dominant bases spanning a space of moments (or derivatives of transfer functions) [4, 10].

There are many model order reductions [2, 13–15] developed for nonlinear circuits as well. Similarly, there can be a nonlinear mapping def ned by a function ϕ

$$z = \phi(x), \quad x = \phi^{-1}(z).$$

For the simplicity of illustration, we assume an ordinary differential equation (ODE) form below

$$\dot{x} = f(x,t) + \mathcal{B}u(t)$$

for the DAE in (3). Since

$$\dot{z} = \frac{d\phi}{dx}\frac{dx}{dt} = (\frac{d\phi}{dx}f(x,t)) + (\frac{d\phi}{dx}\mathcal{B})u(t),$$

we have

$$\dot{z} = \hat{f}(z,t) + \hat{\mathcal{B}}u(t),\ \hat{f}(z,t) = \left[\frac{d\phi}{dx}f(x,t)\right]\big|_{x=\phi^{-1}(z)},\ \hat{\mathcal{B}} = \frac{d\phi}{dx}\mathcal{B}.\qquad(4)$$

As such, if we can f nd a lower-dimensioned mapping function ϕ ($\in N \times q$), the original nonlinear system can be reduced within a tangent subspace spanned by $d\phi/dx$ (or called manifold).

The work in [2] related the above nonlinear mapping function ϕ with a trajectory piece-wise linear (TPWL) method [13]. It leads to a local two-dimensional (2D) projection [2]. Since such a local 2D-projection is constructed from local tangent subspaces, it maintains

Analog Mismatch Analysis by Stochastic Nonlinear Macromodeling   261

a high accuracy. However, it could be computationally expensive to project and store when the number of local tangent subspaces is large. On the other hand, the TPWL method [13] approximated the nonlinear mapping function ϕ by aggregating those local tangent subspaces with the use of a global singular-value-decomposition (SVD). This results in a global one-dimensional (1D) projection. Obviously, the global 1D-projection leads to an efficient projection and runtime. On the other hand, the accuracy of the TPWL model order reduction is limited because the information in the dominant bases of each local tangent subspace are lost during the global SVD [2]. In Section 4, we introduce an incremental aggregation that can balance the speed and accuracy. In addition, we also extend the nonlinear model order reduction to consider the stochastic mismatch.

3. Stochastic Transient Mismatch Analysis

3.1. Stochastic Mismatch Current Model

Directly adding the stochastic mismatch ξ as a parameter into the state variable x of (3) would lead to a difficulty that $f(x,\dot{x},\xi)$ may not be differentiable. Similar to SiSMA [1], we model the mismatch as a current source $\mathbf{i}(x,\xi)$ added at the right-hand-side (rhs) of (3)

$$f(x,\dot{x},t) = \mathcal{F}\,\mathbf{i}(x,\xi) + \mathcal{B}u(t). \tag{5}$$

Here, $\mathcal{F}$ is the topology matrix describing how to connect $\mathbf{i}$ into the circuit.

Based on the BPV equation (2), the stochastic current source $\mathbf{i}$ has a form of

$$\mathbf{i}(x,\xi) = n(x)\sum_l g^\beta(p_l)\xi_l. \tag{6}$$

Here, ξ_l is a random variable associated with a stochastic distribution $W(\xi_l)$ for the parameter p_l. $n(x)$ describes the biasing-dependent condition (depending on $x,\dot{x}$), provided from a nominal transient simulation. $g^\beta(p_l)$ is a constant for the parameter p_l at operating region β. For example, for one CMOS transistor with respect to the parameter area A, ξ_A is one Gaussian random variable, $g^\beta(A)$ is $\kappa^\beta/\sqrt{A}$ and $n(x)$ becomes I_d. In general, $g^\beta(p_l)$ can be either derived based on the analytical device equations or can be practically characterized from measurements [7].

3.2. Perturbation Analysis

Assuming that the impact of the local mismatch is small, (5) can be solved by treating the rhs-term for mismatch as a perturbation to the nominal trajectory $x^{(0)}(t)$ of the circuit. Here, $x^{(0)}(t)$ is the nominal state variable or solution of the nonlinear circuit equation below

$$f(x^{(0)},\dot{x}^{(0)},t) = \mathcal{B}u(t). \tag{7}$$

With a first-order Taylor expansion of $f(x,\dot{x},t)$ in (5), it leads to

$$f(x^{(0)},\dot{x}^{(0)},t) + \frac{\partial f(x,\dot{x},t)}{\partial x}(x-x^{(0)}) + \frac{\partial f(x,\dot{x},t)}{\partial \dot{x}}(\dot{x}-\dot{x}^{0})$$
$$= \mathcal{F}\,\mathbf{i}_n(x^{(0)},\xi) + \mathcal{B}u(t). \tag{8}$$

Or

$$G(x^{(0)}, \dot{x}^{(0)})x_m + C(x^{(0)}, \dot{x}^{(0)})\dot{x}_m = \mathcal{F}\, \mathbf{i}_n(x^{(0)}, \xi), \tag{9}$$

where

$$G(x^{(0)}, \dot{x}^{(0)}) = \frac{\partial f(x, \dot{x}, t)}{\partial x}\Big|_{x=x^{(0)}, \dot{x}=\dot{x}^{(0)}}$$

$$C(x^{(0)}, \dot{x}^{(0)}) = \frac{\partial f(x, \dot{x}, t)}{\partial \dot{x}}\Big|_{x=x^{(0)}, \dot{x}=\dot{x}^{(0)}} \tag{10}$$

are the linearized conductive and capacitive components stamped by the companion models in SPICE, and $x_m = x - x^{(0)}$ is the f rst-order perturbed mismatch response. Recall that $x^{(0)}(t)$ and $\dot{x}^{(0)}(t)$ are a number of time-dependent biasing points along the transient trajectory.

3.3. Non-Monte Carlo Analysis by SOP Expansion

Next, instead of performing the expensive Monte-Carlo or the correlation analysis, the perturbed SDAE (9) with the random variable ξ is solved through an expansion of the stochastic orthogonal polynomial (SOP) [17]. Different random processes are related to the different orthogonal polynomial. The 'Homogeneous Chaos' can be used as the span of Hermite polynomial functionals for a Gaussian process [17]. In this chapter, we assume that the random process parameters for the local mismatch have a Gaussian distribution. Therefore, an according Hermite polynomial (one variable)

$$\Phi(\xi) = [\Phi_1(\xi), \Phi_2(\xi), \Phi_3(\xi), ...,]^T = [1, \xi, \xi^2 - 1, ...,]^T \tag{11}$$

is used to construct the expansion basis to calculate the mean and the variance of $x_m(t)$.

The stochastic state variable $x_m(t)$ is f rst expanded by

$$x_m(t) = \sum_i \alpha_i(t)\Phi_i(\xi). \tag{12}$$

Note that for different random processes, many other orthogonal polynomials can be selected as well based on a so-called *Askey scheme* [17].

Then, when applying the inner-product of the residue error

$$\begin{aligned} \Delta(\xi) \;=\;& G(x^{(0)}, \dot{x}^{(0)}) \sum_i \alpha_i(t)\Phi_i(\xi) + C(x^{(0)}, \dot{x}^{(0)}) \sum_i \dot{\alpha}_i(t)\Phi_i(\xi) \\ &-\; \mathcal{F}\, n(x^{(0)}) \sum_l g^\beta(p_l)\xi_l \end{aligned}$$

by the orthogonal basis $\Phi_j(\xi)$, it results in

$$< \Delta(\xi), \Phi_j(\xi) > = \int_\xi \Delta(\xi)\Phi_j(\xi)W(\xi)d\xi = 0. \tag{13}$$

Here, $W(\xi)$ is the probability distribution of the random variable ξ. We assumed a Gaussian distribution of $W(\xi)$ for all parameters here.

Without the loss of generality, for one random variable ξ of one geometrical parameter p, it is easy to verify that (13) leads to

$$\alpha_0 = 0, \quad \alpha_2 = 0$$
$$G(x^{(0)}, \dot{x}^{(0)})\alpha_1(t) + C(x^{(0)}, \dot{x}^{(0)})\dot{\alpha}_1(t) = \mathcal{F}\, n(x^{(0)})g^\beta(p) \tag{14}$$

with a second-order expansion of $x_m(\xi)$. The according standard-deviation is thereby given by

$$Var < x_m(\xi) >= \alpha_1^2 Var(\xi) + \alpha_2^2 Var(\xi^2 - 1) = \alpha_1^2.$$

The first-order SOP coefficient $\alpha_1(t)$ in (14) can be solved by a Backward-Euler integration

$$(G_k + \frac{1}{h}C_k)\alpha_1(t_k) = \frac{1}{h}C_k\alpha_1(t_k - h) + \mathcal{F}\, \mathbf{i}_k \tag{15}$$

where

$$G_k = G(x_k^{(0)}, \dot{x}_k^{(0)}), \quad C_i = C(x_k^{(0)}, \dot{x}_k^{(0)}), \quad \mathbf{i}_k = n(x_k)\sum_l g^\beta(p_l)$$

are Jacobians and the mismatch current-source at the k-th time-instant along the nominal trajectory $x^{(0)}$.

Clearly, a native application of the above perturbation-based mismatch analysis is still slow, since G_k, C_k and $\mathbf{i}_k$ have to be evaluated during every time-step along the nominal trajectory. Instead of linearizing along the full nominal trajectory, in Section 4, only K snapshots along the nominal trajectory are used in the frame of a macromodeling.

3.4. One CMOS Transistor Example

For one CMOS transistor with a geometric parameter A and the according Gaussian random variable ξ_A, (14) becomes

$$(G_k + \frac{1}{h}C_k)\alpha_1(t_k) = \frac{1}{h}C_k\alpha_1(t_k - h) + \frac{\kappa^\beta}{\sqrt{A}} \cdot (I_d)_k$$

at the k-th time-step. Recall that G_k, C_k and $(I_d)_k$ are the nominal conductance (g_{ds}), capacitance (c_{ds}) and channel current I_d evaluated at t_k, $g^\beta(A)$ is $\kappa^\beta/\sqrt{A}$ and $n(x)$ becomes I_d. Note that κ^β is the extracted constant from Pelgrom's model.

As such, the transient mismatch voltage ($x_m = \alpha_1(t)\Phi_1(\xi_A)$) of this transistor has a time-varying standard variance $\alpha_1(t)^2$, solved from the above perturbation equation. Usually, $\kappa^\beta/\sqrt{A}$ is about few percentages of the nominal channel current I_d. More importantly, for thousands of different typed transistors, we can simultaneously solve the transient mismatch vector using (14) with a generally characterized $g^\beta(p_l)$ by the BPV model [7].

4. Macromodeling for Mismatch

Instead of performing a full simulation for the nominal transient and transient mismatch, we can take K snapshots along a nominal transient trajectory and find subspaces, or macromodels, from the K snapshots with respect to right-hand-sides of the nominal input and stochastic current-source, respectively. Afterwards, eff cient transient and transient mismatch analysis can be performed along the full transient trajectory using macromodels. In the following, for the nominal transient, we f rst introduce an incremental TPWL method to balance the accuracy and eff ciency when generating the macromodel. We then extend this approach to handle the stochastic mismatch.

4.1. Incremental Trajectory-Piecewise-Linear Modeling

4.1.1. Local Tangent Subspace

Given a set of typical inputs, we take K snapshots $\{x_1^{(0)}, ..., x_K^{(0)}\}$ along a nominal transient trajectory $x^{(0)}(t)$, and linearize the DAE (3) at K snapshots (or biasing points), with the f rst snapshot x_1 taken at the ic point. The linearized DAE at k-th ($k = 1, ..., K$) snapshot is

$$G_k(x - x_k^{(0)}) + C_k(\dot{x} - \dot{x}_k^{(0)}) = \delta_k, \; \delta_k = \mathcal{B}u(t_k) - f(x_k^{(0)}, \dot{x}_k^{(0)}, t_k)$$

where δ_k represents the rhs-source and the 'non-equilibrium' update. In frequency domain, $x_k^{(0)}$ at the k-th snapshot is contained by a subspace of moments $\{A_k, A_k R_k, A_k^2 R_k, ...,\}$ expanded at a frequency-point s_0, where

$$A_k = (G_k + s_0 C_k)^{-1} C_k, \; R_k = (G_k + s_0 C_k)^{-1} \delta_k$$

are two moments matrices.

With the use of a block-Arnoldi orthonormalization [10], a q'-th order projection matrix V_k ($\in N \times q'$) with q' bases

$$V_k = [v_k^1, v_k^2, ..., v_k^{q'}]$$

can be constructed locally to represent that local subspace. We call v_k^i ($k = 1, ..., K; i = 1, ..., q'$) as the f rst-$q'$ *dominant bases* of one V_k. Here, for each v_k^i the subscript describes the index of the local subspace, and the superscript describes the index of the order of the dominant base. It f nds a linear coordinate transformation V_k that maintains $||z - z_k^{(0)}|| \approx ||x - x_k^{(0)}||$. Moreover, as discussed in the following part, those V_ks span a subspace for $d\phi/dx$, the tangent (or manifold) of the mapping function ϕ introduced in Section 2. As such, we call the space spanned by V_ks as *local tangent subspace*.

4.1.2. Local and Global Projection

As discovered in [2], one approach to approximate the nonlinear mapping function ϕ introduced in Section 2 is

$$x = \phi^{-1}(z) \approx \sum_{k=1}^{K} w_k [x_k + V_k(z - z_k^{(0)})] \tag{16}$$

and

$$z = \phi(x) \approx \sum_{k=1}^{K} w_k [z_k + V_k^T(x - x_k^{(0)})] \tag{17}$$

Analog Mismatch Analysis by Stochastic Nonlinear Macromodeling   265

where w_k ($\sum_{k=1}^{K} w_k = 1$) is the weighted kernel function, which depends on the distance between a point on the trajectory and a linearization point [13].

Based on the equations (4), (16) and (17), a nonlinear model order reduction is derived in terms of a local two-dimensional (2D) projection

$$\sum_{l=1}^{K}\sum_{k=1}^{K} w_l w_k \left[V_l^T G_k V_k (z - z_k^{(0)}) + V_l^T C_k V_k (\dot{z} - \dot{z}_k^{(0)}) \right] = \sum_{l=1}^{K} w_l V_l^T \delta_k, \qquad (18)$$

where we assume that all V_ks are reduced to the same order q'. For circuits with a sharp transition (input) or strong nonlinearity (device), the number of sampled snapshots is required to be large to maintain a high accuracy. For such kind of circuits, our experiments show that the number of sampled snapshots (or neighbors) has to be large to produce a good accuracy. As such, the computational cost would be prohibitive by the local 2D-projection (18) in [2].

On the other hand, the TPWL method in [13] approximates the nonlinear mapping function ϕ by aggregating the local subspace V_k ($\in N \times q'$) into an unif ed global subspace $span\{V_1, V_2, ..., V_K\}$, which is further compressed into a lower-dimensioned subspace $\mathcal{V}$ ($\in N \times q, q << N$) by a singular-value-decomposition (SVD)

$$\mathcal{V} = SVD_q ([V_1, V_2, ..., V_K]).$$

We call this procedure as *global aggregation*. A global aggregation results in a global one-dimensional (1D) projection by

$$\sum_{k=1}^{K} w_k \left[\mathcal{V}^T G_k \mathcal{V} (z - z_k^{(0)}) + \mathcal{V}^T C_k \mathcal{V} (\dot{z} - \dot{z}_k^{(0)}) \right] = \sum_{k=1}^{K} w_k \mathcal{V}^T \delta_k. \qquad (19)$$

Clearly, such a global 1D-projection has a smaller projection time and storage than the local 2D-projection. However, at the same time, since the dominant bases of those local V_ks are interpolated by the global aggregation, the global 1D-projection usually requires a higher-order q to achieve an accuracy similar to the local 2D-projection with the order q' ($q' < q$) [2].

4.1.3. Incremental Aggregation of Subspaces

The local 2D-projection (18) requires a longer runtime and larger storage compared to the global 1D-projection (19). On the other hand, the global 1D-projection (19) by $\mathcal{V}$ is less accurate than the local 2D-projection (18). As such, we need a procedure that can balance both of the accuracy and eff ciency.

The $d\phi/dx$ (manifold) is covered by those local tangent subspaces $\{V_1, V_2, ..., V_K\}$ along the trajectory, where each V_k is further composed of different orders of dominant bases: $\{v_k^1, v_k^2, ..., v_k^{q'}\}$. An effective aggregation thereby needs to consider the order or the dominance of those bases. This motivates us to f rst decompose the space spanned by those local tangent subspaces according to the order. As such, (16) becomes

$$x = \phi^{-1}(z) \approx \sum_{k=1}^{K} w_k x_k + \sum_{k=1}^{K} w_k \sum_{p=1}^{q'} v_k^p (z - z_k^{(0)})$$

$$= \sum_{k=1}^{K} w_k x_k + \sum_{p=1}^{q'} \sum_{k=1}^{K} v_k^p w_k (z - z_k^{(0)})$$

$$= \sum_{k=1}^{K} w_k x_k + \left[v_1^1 w_1 (z - z_1^{(0)}) + \dots + v_K^1 w_K (z - z_K^{(0)}) \right]$$

$$+ \dots + \left[v_1^q w_1 (z - z_1^{(0)}) + \dots + v_K^q w_K (z - z_K^{(0)}) \right]. \tag{20}$$

Therefore, we can form a *global tangent subspace* in the order of the dominant bases by

$$span\{v_1^1, v_2^1, ..., v_K^1\}, \ ..., \ span\{v_1^{q'}, v_2^{q'}, ..., v_K^{q'}\}.$$

A global projection matrix $\mathcal{V}$ is accordingly constructed below in a fashion of an *incremental aggregation*. We first aggregate each global tangent subspace by orders

$$\mathbf{V}_1 = SVD_q \left([v_1^1, ..., v_K^1] \right), \ ..., \ \mathbf{V}_{q'} = SVD_q \left([v_1^{q'}, ..., v_K^{q'}] \right). \tag{21}$$

I.e., we identify a $\mathbf{V}_p$ $(p = 1, ..., q')$ to represent the p-th order global tangent subspace.

Then, the global projection matrix $\mathcal{V}$ is further aggregated

$$\mathcal{V} = SVD_q \left([\mathbf{V}_1, \mathbf{V}_2, ..., \mathbf{V}_{q'}] \right) \tag{22}$$

by those global tangent subspace in an order of dominance. As shown by experiments, usually we can choose a much lower q' ($q' \ll q$) for each local tangent subspace V_k, and the order q depends on the number of snapshots. For circuits with the sharp transition (input) or strong nonlinearity (device), the number of snapshots is large and so does q.

Clearly, as the local tangent subspace is incrementally aggregated according to their ordered bases, the information of those dominant bases at low orders are preserved. As shown by experiments, when compared to the previous TPWL method [13], this incremental aggregation results in a better accuracy yet with a similar computational cost in the projection time and memory storage. Moreover, our incremental aggregation can also consider more sampled biasing (linearizion) points than the approach in [2], whereas the computational cost of of the local 2D-projection would increase dramatically.

4.2. Stochastic Extension for Mismatch

We further extend the above discussion to build the TPWL macromodel for the stochastic mismatch. Instead of linearizing the DAE (3), we linearize the SDAE (14) at K snapshots along the nominal trajectory similarly, and construct the local tangent subspace V_k by

$$A_k' = (G_k + s_0 C_k)^{-1} C_k, \ R_k' = (G_k + s_0 C_k)^{-1} \delta_k'$$

where δ_k' is determined by the non-equilibrium correction associated with $\mathcal{F} \mathbf{i}_k$. Then, we can build the similar incrementally aggregated mapping $\mathcal{V}$ through (21) and (22).

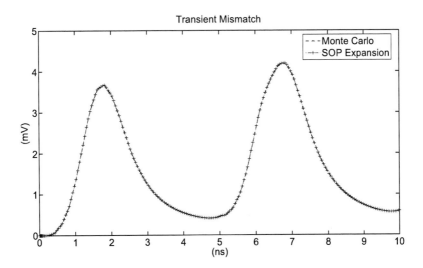

Figure 1. Transient mismatch (the time-varying standard deviation) comparison at output of a bjt-mixer with distributed inductor: the exact by Monte-Carlo and the exact by SOP-expansion.

Afterwards, the global macromodel is constructed from a set of weighted local macromodels by

$$\sum_{k=1}^{K} w_k \cdot \left[\mathcal{V}^T G_k \mathcal{V} \alpha_1(t) + \mathcal{V}^T C_k \mathcal{V} \dot{\alpha}_1(t) - \mathcal{V}^T \mathcal{F} \cdot \mathbf{i}_k \right] = 0 \qquad (23)$$

to calculate the transient mismatch. We call such a macromodeling as *isTPWL method*. Using such a macromodel sampled from K snapshots, we can then efficiently perform a transient mismatch analysis for the full trajectory.

5. Numerical Results and Analysis

The proposed mismatch algorithm has been implemented in C and Matlab. A modernized Spice3 (http://ngspice.sourceforge.net/) is used to generate the K snapshots of a nominal trajectory and to extract the mismatch current model. The SOP expansion, Backward-Euler, and incremental and stochastic TPWL (isTPWL) are implemented in Matlab. For the comparison, the TPWL method and maniMOR method are implemented exactly following the procedure described in [13] and [2], respectively. For example, as for the TPWL method [13], the state variables at snapshots are added to have a 'richer' information during the global aggregation. The flow under the Monte-Carlo analysis is also implemented as the baseline with 1000 iterations. All experimental results are measured on an Intel duel-core 2.0GHZ PC with 2GB memory. We compared the accuracy and study the scalability of our method with four industrial analog/RF circuits. They contain different transistors such as diodes, BJTs and CMOSs. The circuits also include the extracted parasitics so the matrix time is dominant. For the characterization of $g^\beta(p_l)$, we use Pelgrom's model for CMOS transistors and BPV model for diodes and BJTS, all resulted in $\sim 10\%$ variation

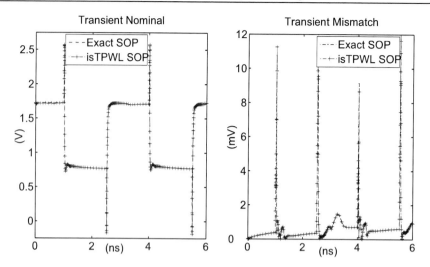

Figure 2. Transient nominal ($x^{(0)}(t)$) (a) and transient mismatch ($\alpha_1(t)$) (b) for one output of a coms-comparator by the exact SOP and the isTPWL.

Table 1. Scalability comparison of runtime and error for the exact model with MC, the exact model with SOP and the isTPWL macromodel with SOP

1	2	3	4	5	6	7	8	9	10
Circuit	# of nodes	# of steps	# of snapshots	# of orders	MC time(s)	Exact SOP time(s)	error	SOP+isTPWL time(s)	error
diode chain	802	225	24	25	520.1	0.53	0.41%	0.02	0.43%
BJT mixer-1	238	135	25	25	338.0	0.34	0.29%	0.02	0.36%
BJT mixer-2	1248	219	83	45	348.0	0.20	0.18%	0.04	0.24%
CMOS comp.	654	228	75	60	412.1	0.39	0.41%	0.08	0.62%

from the nominal bias $n(x)$ (for example, I_d for CMOS transistor). In addition, we measure the waveform error by taking the averaged difference of two waveforms. Three waveforms are measured at each time-step: one is the transient nominal ($x^{(0)}(t)$), the other is the transient mismatch ($\alpha_1(t)$, the time-varying standard deviation), and the last one is the transient ($x(t)$, the nominal plus the standard deviation).

5.1. Comparison of Mismatch Waveform-Error and Runtime

We first compare the waveform accuracy of the transient mismatch between the MC method (1000 iterations) and the exact SOP expansion, and further compare the accuracy with the isTPWL macromodel. In addition, we also compare the waveform of the transient mismatch and the waveform by adding mismatch as one initial condition as SiSMA [1] does. Finally, we summarize the runtime and waveform error in a table.

The first example is a BJT-mixer circuit including an extracted distributed inductor with 238 state variables. We compare the waveforms by solving the perturbed SDAE (9) with the Monte-Carlo (MC) analysis and the SOP expansion, respectively. We apply a MC with Gaussian distribution 1000 times at one time-step and calculate the time-varying standard

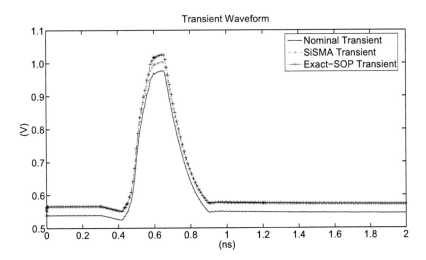

Figure 3. Transient waveform comparison at output of a diode-chain: the transient nominal, the transient with mismatch by SiSMA (adding mismatch at *ic* only), the transient with mismatch by our method (adding mismatch at transient trajectory).

deviation. It takes 348 seconds for the transient mismatch by the MC analysis, and 0.20 second (more than 1000 times speedup) for the exact SOP expansion up to the second order with error less than 0.18%. Clearly, as shown in Fig.1, the two waveforms of transient mismatches are virtually identical.

Next, we show a further speed improvement by macromodeling. The second example is a CMOS-comparator including an extracted power supply with 654 state variables. We compare waveforms of the exact SOP expansion and the one further reduced by isTPWL. Fig.2 (a) shows the comparison of the transient nominal, and Fig.2 (b) shows the comparison of the transient mismatch. 75 snapshots are used to generate the macromodel, and the original model is reduced to a macromodel with the order of 60. For a short-transient with 228 time-steps, it takes 0.39s for the exact and 0.08s for the isTPWL (5 times speedup). The waveforms error by isTPWL is 0.62%.

We further compare the transient mismatch waveforms on how to add the mismatch. The first is to add the stochastic mismatch only as the *ic* condition, the procedure used in SiSMA [1]. The second is our approach by adding the stochastic mismatch during every time-step. A Diode-chain with 802 state variables is used. Figure 3 shows one waveform of the transient nominal, and two waveforms with mismatches added differently. Clearly, the waveform with mismatch added at *ic* shows a non-negligible difference.

Finally, we summarize the runtime and error of four different analog/RF circuits in Table 1. The waveform error here is defined as the relative difference between the exact and the macromodel. The runtime here is the total simulation time. Column 1–5 summarize the circuit type, size, time-steps in transient simulation, snapshot points, and reduced order. Column 6 shows the runtime of MC by 1000 iterations, Column 7–8 show the runtime to simulate the exact model with the SOP expansion and the error compared to MC, Column 9–10 show the runtime and error of macromodels reduced by isTPWL. We find that the

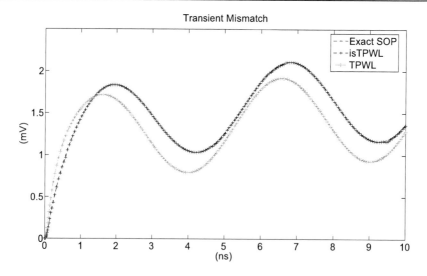

Figure 4. Transient mismatch ($\alpha_1(t)$, the time-varying standard deviation) comparison at output of a BJT-mixer with distributed substrate: the exact by SOP-expansion, the macromodel by TPWL (order 45), and the macromodel by isTPWL (order 45). The waveform by isTPWL is visually identical to the exact SOP.

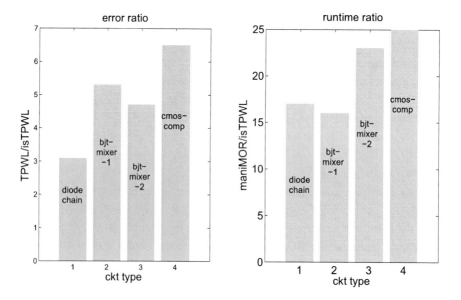

Figure 5. (a) Comparison of the ratio of the waveform error by TPWL and by isTPWL under the same reduction order; (b) Comparison of the ration of the reduction runtime by maniMOR and by isTPWL under the same reduction order. In both cases, isTPWL are used as the baseline.

SOP expansion reduces the runtime by 1000 times yet with an error of 0.23% on average. Moreover, the macromodel by isTPWL further reduces the runtime up to 25 times (diode chain) yet with an error up to 0.43%. This demonstrates the eff ciency and accuracy of our isTPWL method for the transient mismatch analysis.

5.2. Comparison of TPWL Macromodel

In this part of experiment, we further show the accuracy and runtime improvement by isTPWL. First, Fig.4 presents the transient-mismatch waveform comparison for a BJT-mixer including the distributed substrate with total 1248 state variables. 83 snapshots are used and both TPWL and isTPWL reduce the original model to a macromodel with the order of 45. We f nd that the waveform by isTPWL is visually identical to the exact SOP but the waveform by TPWL [13] shows a non-negligible waveform error 4.5 times larger than the one by our isTPWL. Fig.5 further summarizes the comparison by the four circuits used before. Fig.5 (a) is the comparison of the ratio (TPWL vs. isTPWL) of waveform-errors for simulated macromodels by TPWL [13] and by isTPWL under the same reduction order. Fig.5 (b) is the comparison of the ratio (maniMOR vs. isTPWL) of the reduction time for reduced macromodels by maniMOR [2] and by isTPWL under the same reduction order. In both cases, isTPWL are used as the baseline when calculating the ratio. We f nd that our isTPWL method is 5 times more accurate than TPWL [13] and is 20 times faster than maniMOR [2] on average. This clearly demonstrates the advantage to use the incremental aggregation.

6. Conclusion

This chapter has presented a fast non-Monte-Carlo mismatch analysis. It models the mismatch by a current source associated with a random variable and forms a stochastic differential-algebra-equation (SDAE). The random variable in SDAE is expanded by stochastic orthogonal polynomials. This leads to an eff cient solution without using the Monte-Carlo or correlation analysis. Moreover, the SDAE has been solved by an improved trajectory-piecewise-linear (TPWL) mode order reduction, called isTPWL. An incremental aggregation has been introduced to balance the eff ciency and accuracy when generating the macromodel. Numerical results show that when compared to the Monte-Carlo method, the presented method is 1000 times faster with a similar accuracy. Moreover, on average the isTPWL method is 5 times more accurate than the work in [13] and is 20 times faster than the work in [2]. In addition, the use of a reduced macromodel reduces the runtime by up to 25 times when compared to the use of a full model.

Acknowledgement

The work is sponsored in part by NSF CAREER Award No. CCF-0448534, in part by NSF grant under No. OISE-0623038 and in part by National Natural Science Foundation of China (NSFC) grant under No. 60828008. The work of Hao Yu is sponsored in part by NRF2010NRF-POC001-001 (Singapore) and MOE ACRF Tier-1 (Singapore).

References

[1] G. Biagetti, S. Orcioni, C. Turchetti, P. Crippa, and M. Alessandrini, "SiSMA: A tool for eff cient analysis of analog cmos integrated circuits affected by device mismatch," *IEEE Trans. on CAD*, pp. 192–207, 2004.

[2] C.J.Gu and J.Roychowdhury, "Model reduction via projection onto nonlinear manifolds, with applications to analog circuits and biochemical systems," in *Proc. IEEE/ACM ICCAD*, 2008.

[3] A. Demir, E. Liu, and A.Sangiovanni-Vincentelli, "Time-domain non-monte carlo noise simulation for nonlinear dynamic circuits with arbitrary excitations," *IEEE Trans. on CAD*, pp. 493–505, 1996.

[4] P. Feldmann and R. W. Freund, "Eff cient linear circuit analysis by pade approximation via the lanczos process," *IEEE Trans. on CAD*, pp. 639–649, 1995.

[5] J. Kim, K. Jones, and M. Horowitz, "Fast, non-monte-carlo estimation of transient performance variation due to device mismatch," in *Proc. ACM/IEEE DAC*, 2007.

[6] H. Masuda, S. Ohkawa, A. Kurokawa, and M. Aoki, "Challenge: Variability characterization and modeling for 65- to 90-nm processes," in *Proc. IEEE CICC*, 2005.

[7] C. McAndrew, J. Bates, R. Ida, and P. Drennan, "Eff cient statistical BJT modeling, why beta is more than ic/ib," in *Proc. IEEE Bipolar/BiCMOS Circuits and Tech. Meeting*, 1997.

[8] N. Mi, J. Fan, S. X.-D. Tan, Y. Cai, and X. Hong, "Statistical analysis of on-chip power delivery networks considering lognormal leakage current variations with spatial correlations," *IEEE Trans. on CAS-I*, vol. 55, no. 7, pp. 2064–2075, Aug. 2008.

[9] N. Mi, S. X.-D. Tan, Y. Cai, and X. Hong, "Fast variational analysis of on-chip power grids by stochastic extended krylov subspace method," *IEEE Trans. on CAD*, vol. 27, no. 11, pp. 1996–2006, 2008.

[10] A. Odabasioglu, M. Celik, and L. Pileggi, "PRIMA: Passive reduced-order interconnect macro-modeling algorithm," *IEEE Trans. on CAD*, pp. 645–654, 1998.

[11] J. Oehm and K. Schumacher, "Quality assurance and upgrade of analog characteristics by fast mismatch analysis option in network analysis environment," *IEEE JSSC*, pp. 865–871, 1993.

[12] M. Pelgrom, A. Duinmaijer, and A. Welbers, "Matching properties of mos transistors," *IEEE JSSC*, pp. 1433–1439, 1989.

[13] M. Rewienski and J. White, "A trajectory piecewise-linear approach to model order reduction and fast simulation of nonlinear circuits and micromachined devices," *IEEE Trans. on CAD*, pp. 155–170, 2003.

[14] J. Roychowdhury, "Reduced-order modelling of time-varying systems," in *Proc. AS-PDAC*, 1999.

[15] S.K.Tiwary and R.A.Rutenbar, "Faster, parametric trajectory-based macromodels via localized linear reductions," in *Proc. IEEE/ACM ICCAD*, 2006.

[16] S. Vrudhula, J. M. Wang, and P. Ghanta, "Hermite polynomial based interconnect analysis in the presence of process variations," *IEEE Trans. on CAD*, pp. 2001–2011, 2006.

[17] D. Xiu and G. Karniadakis, "The Wiener-Askey polynomial chaos expansion for stochastic differential equations," *SIAM J. Scientific Computing*, pp. 619–644, 2002.

INDEX

A

Abraham, 146, 149, 150, 152, 154, 156, 158, 160, 162, 164, 166, 168
accelerometers, 42
actuators, 42
adaptation, 235, 236, 238, 244, 253
ADC, 42, 259
advancement, 4, 12, 17, 41
aggregation, 257, 258, 261, 265, 266, 267, 271
algorithm, 12, 13, 14, 15, 16, 31, 38, 55, 125, 153, 154, 155, 160, 163, 164, 165, 166, 173, 174, 175, 177, 180, 181, 182, 184, 189, 191, 197, 198, 200, 204, 206, 208, 209, 222, 224, 226, 227, 229, 233, 235, 236, 239, 242, 243, 244, 247, 251, 253, 254, 267, 272
amplitude, 30, 42, 43, 141
analog signal processing, vii, 69, 118, 121
animal behavior, 236
artificial intelligence, 172, 173
Asia, 37, 38, 39, 40, 231, 233
automation, vii, 4, 5, 32, 36, 121, 122, 124, 126, 133, 142, 146, 173

B

bandgap, 47, 48, 49, 50, 67
bandwidth, 6, 7, 27, 41, 58, 60, 61, 65, 126, 219, 245, 249
base, 11, 17, 47, 48, 49, 50, 51, 52, 53, 56, 58, 61, 65, 66, 67, 236, 259, 264
behaviors, 131
benchmarks, 20
benefits, 42
bias, 51, 55, 56, 63, 64, 67, 123, 125, 159, 245, 249, 268
binary decision, 3, 11, 13, 14, 16, 36
birds, 221
bounds, 218
Brazil, 40
building blocks, 118

C

Cairo, 93
calibration, 53, 54
case studies, 3, 232, 253
cell phones, 42, 46
chaos, 121, 122, 127, 129, 133, 134, 137, 138, 141, 142, 146, 273
chaotic masking, 141
chemical, 236
China, 3, 39, 271
chromosome, 123, 125, 126
clarity, 165, 171
classes, 10
classical methods, 9
classification, 117, 154, 216
CNN, 150, 153, 154, 155, 163, 166
cognition, 222
commercial, 29
communication, 46, 66, 121, 122, 141, 142, 147
communication systems, 46, 122, 142
community, 7
compatibility, 46
compensation, 4, 33, 37, 245
complex numbers, 23, 24
complexity, 5, 9, 13, 17, 18, 19, 38, 69, 70, 77, 154, 155, 156, 166, 171, 172, 177, 178, 180, 188, 189, 203, 209, 213, 215, 218, 258
compliance, 172
composition, 50
compression, 154
computation, 3, 5, 7, 11, 22, 23, 31, 34, 35, 36, 40, 83, 84, 150, 217, 239
computer, 9, 10, 11, 20, 36, 44, 45, 46, 83, 149, 150, 172, 174, 175, 176, 177, 181, 186, 187, 188, 189, 200
computer use, 36
computer-aided design (CAD), 3, 7, 38, 42, 43, 172, 173, 210, 272, 273
computing, 17, 153, 162, 171, 173, 181, 186, 199, 200, 219, 232, 239, 241, 246, 253
conception, 152
concreteness, 179
conditioning, 69

conductance, 8, 263
conductivity, 185, 190
configuration, 60, 62, 64, 135, 138, 141
connectivity, 153
construction, 5, 6, 11, 13, 16, 17, 18, 19, 20, 22, 36, 127, 145, 184, 196, 209
consumption, 36, 149, 150, 154, 162, 166, 219
contour, 175
controlled voltage source (VS), 8
convergence, 137, 204, 209, 219, 235
correlation, 22, 208, 209, 225, 262, 271
correlation analysis, 262, 271
correlations, 272
cost, 23, 36, 41, 47, 58, 66, 171, 174, 177, 205, 213, 225, 226, 246, 247, 258, 265, 266
CPU, 125, 171, 173, 174, 177, 178, 180, 186, 194, 195, 200, 203, 204, 206, 207, 208, 209
creative process, 43
CVD, 67

D

damping, 30
data structure, 10, 11, 32, 36
decomposition, 13, 171, 172
degradation, 220
Delta, 233
dependent variable, 173, 175, 183, 186, 190, 195, 198
deposits, 223
derivatives, 10, 70, 72, 182, 183, 205, 208, 260
designers, 5, 6, 7, 22, 28, 214, 220, 235, 258
detectable, 137
detection, 149, 150, 165, 166
deviation, 87, 269
differential equations, 182, 184, 192, 273
diffusion, 51
diffusivity, 51
dilation, 152, 153, 154, 160, 163, 164, 166
diodes, 267
disclosure, 29
discrete variable, 238, 239
discretization, 177
dislocation, 30
distribution, 6, 163, 164, 229, 261, 262, 268
diversity, 164
DOI, 144, 145
dominance, 220, 265, 266
doping, 47, 49, 51, 58, 61, 67
dynamic systems, 204
dynamical systems, 135

E

Egypt, 93
election, 150, 153
electric field, 49, 50, 52
electromagnetic, 214

electron, 47, 51, 67
electronic circuits, 130, 171, 172, 173, 174, 175, 176, 177, 189, 204, 209, 235
electronic systems, 209
electrons, 47, 49, 50, 51
employment, 159
encoding, 122, 123, 142
encryption, 142
energy, 48, 67, 135, 136
environment, 31, 65, 221, 236, 272
equilibrium, 175, 205
erosion, 152, 153, 154, 160, 163, 164, 166
evaporation, 238
evolution, vii, 53, 122, 183, 221, 235
evolutionary computation, 232, 233, 253
extraction, 7, 37, 55, 66, 154, 258

F

fabrication, 41, 44, 45, 214
FFT, 137, 139, 141
filters, 4, 67, 69, 72, 75, 77, 78, 79, 80, 82, 83, 85, 86, 87, 88, 93, 117, 118, 121, 145, 147, 149, 150, 152, 154, 164, 166, 213
fish, 221
fitness, 125, 221, 223
flatness, 72, 86
flights, 222, 223
fluctuations, 159
formation, 65
formula, 7, 26, 27, 33, 85, 180, 190, 191, 195, 204
freedom, 44, 85, 204

G

genes, 123, 125
geometric programming, 244, 247
geometrical parameters, 214, 259
geometry, 42, 44, 259
germanium, 66
grading, 47, 49, 50, 52
graph, 3, 8, 9, 12, 13, 14, 15, 18, 37, 38, 150, 236, 238, 239, 254
grids, 272

H

Hamiltonian, 122, 134, 135, 136, 137, 141, 142, 147
hardness, 41
Heterojunction Bipolar Transistors (HBTs), vii, 41
histogram, 163
hybrid, 152, 231, 232, 253

I

IAM, 273
ICC, 63
identity, 11, 23, 25, 137, 156, 158, 183
illumination, 165, 166
image, 149, 150, 153, 160, 163, 164, 165, 166
images, vii, 150, 153, 154, 155
improvements, 12, 156, 258
independent variable, 54, 173, 174, 175, 176, 179, 180, 183, 184, 186, 190, 191, 202, 203
individuals, 125, 221, 222
inductor, 58, 63, 64, 109, 111, 213, 214, 215, 216, 218, 225, 226, 227, 228, 230, 231, 267, 268
industry, 41
inequality, 218, 219, 220
inertia, 222
information sharing, 221
integrated circuits, 3, 4, 9, 18, 37, 38, 39, 44, 67, 122, 146, 171, 231, 253, 272
integration, 41, 66, 127, 184, 185, 187, 188, 195, 197, 213, 245, 260, 263
intelligence, 254
interface, 7, 31, 42
issues, vii, 41, 42, 66, 67, 150
iteration, 33, 35, 172, 174, 175, 179, 214, 223, 238, 239

J

Japan, 37, 38, 39

L

Lagrange multipliers, 191
latency, 42
laws, 173, 176, 209
lead, 17, 18, 23, 177, 185, 186, 192, 201, 221, 258, 261
leakage, 272
linear function, 84, 85
linear model, 55
linear systems, 182
localization, 22, 39
LTA, 149, 154, 155, 156, 158, 159, 160, 161, 162, 163, 164, 166
Lyapunov function, 172, 204, 205, 206, 207, 208, 209

M

magnitude, 6, 24, 27, 29, 56, 70, 71, 72, 73, 76, 77, 81, 82, 83, 90, 184, 208, 258
majority, 121, 125, 237
Malaysia, 40
manifolds, 272

manipulation, 10, 37, 38, 53
mapping, 19, 25, 260, 261, 264, 265, 266
masking, 141, 142
matrix, vii, 11, 16, 17, 18, 19, 20, 21, 33, 53, 93, 94, 96, 99, 118, 136, 137, 172, 182, 183, 258, 259, 260, 261, 264, 266, 267
measurements, 53, 55, 67, 201, 231, 261
memory, 11, 16, 20, 36, 150, 221, 223, 266, 267
metal-oxide-semiconductor, 122
methodology, 37, 122, 171, 181, 189, 191, 209, 214, 225, 226
microelectronics, 255
MIP, 150, 153
modelling, 214, 273
models, 17, 18, 43, 45, 46, 63, 144, 171, 173, 180, 190, 213, 214, 215, 230, 259, 262, 267, 271
modifications, 32, 166
modules, 41, 149, 150
mole, 49
Morocco, 235, 254
morphology, 150, 152
Moscow, 211
mutation, 221

N

nanometer, 257
Netherlands, 39
network elements, 117
neural network, 150
neutral, 49, 50, 51, 52
next generation, 58
nodes, 8, 10, 19, 96, 123, 127, 172, 184, 185, 186, 188, 189, 195, 196, 268
nonlinear dynamics, 39
nonlinear systems, 154
numerical analysis, 191

O

operational amplifiers, 4, 145
operations, 10, 11, 12, 13, 14, 15, 17, 23, 32, 38, 124, 146, 151, 152, 153, 156, 158, 163, 166, 178, 180
optimal performance, 229
optimization, vii, 3, 5, 6, 22, 24, 31, 32, 33, 34, 35, 36, 55, 66, 67, 171, 172, 173, 174, 175, 176, 177, 178, 179, 180, 181, 182, 183, 184, 185, 186, 187, 188, 189, 191, 192, 194, 195, 196, 197, 200, 201, 205, 206, 208, 209, 210, 211, 214, 218, 219, 220, 221, 222, 223, 226, 227, 228, 229, 230, 231, 232, 233, 235, 236, 238, 244, 246, 247, 251, 253, 254
optimization method, 172, 174, 181, 185, 186, 189, 192, 194, 195, 196, 197, 221, 226
orbit, 66
ordinary differential equations, 177, 179
oscillation, 30, 53, 56, 61, 115, 116

P

Pacific, 37, 40
Pakistan, 255
parallel, 13, 44, 149, 150
Pareto, 220, 223, 229, 246
pattern recognition, 154
photodetectors, 155
pitch, 184, 217
platform, 40
polarity, 113, 114
polarization, 249
portraits, 202, 203
Portugal, 213, 231
principles, 5, 9, 42, 233
probability, 203, 237, 238, 241, 262
probability distribution, 262
processing stages, 149, 150
programming, 9, 11, 173
project, 144, 209, 261
propagation, 259

R

radiation, 41
radius, 26, 30, 74
ramp, 26
reconstruction, 135
recovery, 141
redistribution, 171, 181, 209
rejection, 4, 245, 246, 248, 249, 251, 252, 253
reliability, 41, 46
repetitions, 174
replication, 158, 159
requirements, 41, 46, 69, 150, 154, 168, 213, 218, 219
researchers, 5, 7, 67
resistance, 47, 51, 53, 55, 56, 57, 61, 65, 66, 67, 185, 190, 215, 216, 259
resolution, 154, 162, 219
response, 4, 6, 7, 28, 29, 30, 39, 42, 47, 49, 50, 51, 52, 53, 56, 58, 59, 60, 61, 62, 63, 66, 70, 81, 260, 262
restrictions, 150, 178, 179, 185
room temperature, 65
root, 14, 19, 22, 39, 70
routines, 54, 239
rules, 8, 9, 15, 16, 17, 221, 223, 225

S

saturation, 33, 67, 131, 142, 246, 250
scaling, 131, 221
scattering, 53, 54, 55
search space, 127
second generation, 126, 254

secure communication, vii, 127, 141, 142, 147
semiconductor, 17, 18, 31, 32, 33, 35, 66, 67
sensitivity, 3, 5, 12, 17, 22, 23, 24, 25, 26, 27, 28, 29, 30, 31, 32, 33, 34, 35, 36, 39, 40, 64, 69, 155, 156, 259
sensors, 42, 155
sequencing, 11
shape, 58, 165, 166, 177, 204, 225
showing, 43, 135, 239
signals, 41, 42, 43, 45, 64, 84, 121, 122, 129, 141, 149, 150, 151, 155, 158, 162, 166
silhouette, 165
silicon, 41, 44, 66, 213, 214, 217, 230
silicon-germanium (SiGe), vii
simulation, vii, 4, 6, 18, 20, 33, 39, 44, 45, 46, 49, 58, 59, 61, 62, 66, 67, 125, 129, 137, 147, 159, 160, 172, 213, 214, 228, 229, 230, 235, 236, 248, 251, 252, 253, 258, 261, 263, 269, 272
sine wave, 70, 141
Singapore, 37, 257, 271
software, 22, 43, 63
solution, 8, 36, 44, 45, 46, 156, 158, 172, 173, 174, 176, 177, 178, 182, 190, 191, 192, 193, 198, 205, 206, 219, 220, 221, 222, 225, 235, 236, 238, 258, 261, 271
solution space, 236
South Pacific, 231
specifications, 42, 43, 45, 69, 172, 218, 251
stability, 56, 70, 74, 134, 137, 204, 205, 206, 208, 209
standard deviation, 267, 268, 270
standardization, 10
state, 6, 9, 42, 43, 56, 127, 135, 136, 137, 142, 173, 190, 195, 198, 204, 206, 260, 261, 262, 267, 268, 269, 271
storage, 5, 18, 265, 266
stress, 19, 221, 226
structure, 10, 18, 19, 47, 48, 58, 150, 153, 154, 171, 172, 176, 197, 201, 202, 203, 204, 206, 208, 209, 211, 221
structuring, 150, 152, 153, 164, 166
substrate, 44, 54, 214, 215, 217, 270, 271
subtraction, 32, 64, 127, 141
supplier, 64
suppression, 5, 13
symmetry, 153
synchronization, 122, 134, 135, 137, 138, 139, 141
synchronize, 122, 134, 135, 141, 142
synthesis, vii, 5, 10, 11, 13, 39, 44, 69, 117, 118, 122, 124, 145, 146, 147, 231, 232, 253

T

Taiwan, 39
taxonomy, 232, 253
techniques, 3, 4, 5, 12, 22, 36, 42, 45, 54, 55, 69, 171, 172, 210, 214, 218, 219, 220, 221, 232, 236, 253, 254

Index

technologies, 41, 208, 245

technology, 4, 36, 41, 44, 45, 46, 47, 53, 63, 65, 66, 67, 142, 149, 159, 161, 162, 164, 166, 213, 214, 218, 225, 226, 230, 246, 248, 250, 251

temperature, 47, 49, 51, 63, 65, 175, 227, 228

terminals, 66, 122

time use, 20

topology, 55, 57, 122, 125, 126, 127, 128, 141, 142, 173, 175, 260, 261

tracks, 216, 222, 226

trajectory, 26, 171, 177, 178, 181, 183, 189, 190, 193, 194, 195, 196, 197, 200, 201, 203, 204, 206, 208, 257, 258, 260, 261, 262, 263, 264, 265, 266, 267, 269, 272

transducer, 54

transformation, 25, 56, 89, 117, 132, 183, 209, 260, 264

transformations, 26, 42, 69, 90, 118

transistor, 41, 44, 47, 51, 52, 53, 54, 55, 56, 58, 60, 64, 65, 122, 142, 158, 159, 186, 187, 188, 189, 197, 202, 203, 207, 218, 245, 246, 250, 257, 261, 263, 268

transmission, 141, 142, 143

transport, 50, 51, 67

transportation, 49

Turkey, 254

U

Ukraine, 171

USA, 37, 38, 39, 232

USSR, 210

V

variables, 32, 33, 54, 127, 173, 174, 175, 176, 179, 181, 183, 184, 185, 190, 192, 195, 196, 197, 198, 204, 205, 214, 218, 236, 238, 267, 268, 269, 271

variations, 159, 186, 272, 273

vector, 6, 64, 135, 136, 137, 171, 172, 175, 176, 180, 181, 184, 185, 186, 188, 189, 190, 191, 192, 194, 195, 196, 197, 198, 202, 203, 204, 205, 206, 207, 208, 209, 211, 220, 263

velocity, 67, 221, 222, 223

voltage controlled voltage source (VCVS), 8

W

Washington, 166

Y

yield, 41, 51, 172